ELEMENTS OF DYNAMIC OCEANOGRAPHY

Physical processes of sedimentation
J. R. L. Allen

Elements of geographical hydrology
B. J. Knapp

Atmospheric processes
J. D. Hanwell

The maritime dimension
R. P. Barston & P. Birnie (eds)

Sedimentology: process and product
M. R. Leeder

Sedimentary structures
J. D. Collinson & D. B. Thompson

The changing climate
M. J. Ford

Quaternary paleoclimatology
R. S. Bradley

ELEMENTS OF DYNAMIC OCEANOGRAPHY

David Tolmazin

Marine Sciences Institute,
University of Connecticut

Boston
ALLEN & UNWIN
London   Sydney

Allen & Unwin Inc.,
Fifty Cross Street, Winchester, Mass 01890, USA

George Allen & Unwin (Publishers) Ltd,
40 Museum Street, London WC1A 1LU, UK

George Allen & Unwin (Publishers) Ltd,
Park Lane, Hemel Hempstead, Herts HP2 4TE, UK

George Allen & Unwin Australia Pty Ltd,
8 Napier Street, North Sydney, NSW 2060, Australia

First published in 1985

Library of Congress Cataloging in Publication Data

Tolmazin, David, 1933–
    Elements of dynamic oceanography.
Bibliography: p.
Includes index.
1. Oceanography. 2. Ocean circulation. I. Title.
GC201.2.T65 1985   551.47   84-12383
ISBN 0-04-551070-9 (alk. paper)
ISBN 0-04-551071-7 (pbk. : alk. paper)

British Library Cataloguing in Publication Data

Tolmazin, David
    Elements of dynamic oceanography.
1. Ocean currents 2. Ocean waves
I. Title
551.47   GC201.2
ISBN 0-04-551070-9
ISBN 0-04-551071-7 Pbk

Set in 9 on 11 point Melior by
D. P. Media Limited, Hitchin, Hertfordshire
and printed in Great Britain by
Butler & Tanner Ltd, Frome and London

Природа, прислонясь к моим плечам,
объявит свои детские секреты.

И вот тогда — из слез, из темноты,
из бедного невежества былого
друзей моих прекрасные черты
появятся и растворятся снова.

(And Nature, leaning on my shoulders,
will reveal her child-like secrets.

And only then, out of the tears and darkness,
from the poor ignorance of bygone times,
the splendid features of my friends
will appear and once again dissolve.)

BELLA AKHMADULINA

Preface

The ocean evokes the most romantic images of nature. It is the eternally hostile element that has taken a heavy toll for every act of discovery, sometimes in human lives. No wonder there has always been a romantic aura about those who take to the sea, be they pirates, fishermen, sailors, or even oceanographers. Their exploits, and the ocean itself, have provided ample food for thought and poetic inspiration. Clearly, mankind owes much to the ocean for the progress of civilization.

There is more to wresting the ocean's secrets from its depths than simply the excitement of struggling with the elements. It is the thrill of ideas, of discoveries made by scientific analysis of oceanic phenomena. There have been quite a few renowned oceanographers who have never set foot aboard ship. All they did was to use the general laws of fluid behavior and mathematical formulas as tools to study the ocean and to predict events. Amazing 'armchair' discoveries of currents and deep-sea flows, subsequently confirmed by observations at sea, are fascinating. What a scientist feels when uncovering the true behavior of oceanic phenomena in abstract columns of numbers, in long and cumbersome, or sometimes intriguingly simple, mathematical relations, is exhilaration. My objective has been to bring this delightful esthetic pleasure within everyone's reach – the outcome is this book.

It was about twelve years ago when I first recognized the inherent harmony of the theory of currents. I was probably prompted by H. Stommel's famous 'A survey of ocean current theory' (1957) and by subsequent developments. It seemed as though every new idea in that field arose just in time to fill its own pigeonhole, and add a touch to a balanced picture of ocean dynamics. For some reason it was easy for me to peel every new theory of its mathematical shell and lay bare to my audience the physical gist of the matter in simple words, for indeed there is no idea that cannot be expressed in plain language without resort to formulas. I was surprised to find that my verbal descriptions of dynamic processes were perceived by my students as the most interesting elements in my courses of lectures. They provided a good organizing framework for understanding the basic concepts of ocean current theory. Moreover, using the same approach, I managed to impart the fundamentals of ocean dynamics to students of non-oceanographic departments and even to non-science majors.

This book is largely a review of the historical development of ideas in dynamic oceanography with emphasis on the theory of ocean currents. Starting from very simple axioms and general physical laws, models of gradually increasing complexity are presented in a descriptive conceptual manner. An attempt is made to provide a link between old and new ideas. Some old concepts, such as depth of no-motion, no longer used in oceanographic practice, are retained mainly for educational purposes. The oceanic phenomena are linked with processes in the atmosphere to show the role of currents in present and past climates, as well as in projects for climatic modification.

I felt that the barrier separating me from my readers would be mainly one of vocabulary. Therefore I have made a point of employing as few special terms, and, of course, mathematical symbols, as I can, while trying to explain clearly as many facts as possible.

The book is intended mainly for undergraduate students and new graduates taking courses in dynamic oceanography. It will probably also be useful to specialists who in some way deal with the ocean, such as meteorologists, hydrobiologists, fishermen, scientists in general, and students in many fields. An oceanographer may find of interest certain aspects of ocean dynamics that are based on Soviet literature.

In order to suit the disparate needs of this diverse readership, a mixed system of references and bibliographic citations has been used. Numerous references in the body of the text might discourage students, but some indication of additional reading to achieve a better understanding of the subject seemed desirable. In view of this, the references and bibliography are presented in a number of parts. First, there are those indicating the sources of numeric data and graphic materials. These are followed by 'facilitated' reading for the benefit of those who, for some reason, would prefer to learn about the subject from other popular science writers. Finally, textbooks, monographs and articles are suggested for those readers – and, in all modesty, I hope that there will be some of these too – who feel the urge to embark on a thorough study of the present-day science of oceanography.

D. TOLMAZIN

Acknowledgments

In writing this book, I received valuable help from my colleagues at the Marine Sciences Institute, University of Connecticut, who patiently answered my questions and were of great assistance in my efforts to combine teaching with writing the book. Thanks are also due to Mr K. Erastor. The concept originated from contacts with my friends, Soviet oceanographers with whom I had to part company forever when leaving Russia. I am indebted to Professor F. T. Banner and Dr R. Hall for their valuable comments and suggestions which helped to improve the text substantially. I am grateful to Professor W. J. Pierson for encouragement and support. Finally, I owe an immense debt of gratitude to my wife, Macella, and son, Alec, who, although they have not reviewed my manuscript, are fully aware of how much this book means to me.

I would then like to thank the following individuals and organizations who have given me permission to reproduce copyright material (numbers in parentheses refer to text figures):

Figures 1.2, 5.17 and 8.10 reproduced from *Understanding climatic changes* (1975) with the permission of the National Academy of Sciences, Washington, DC; Figures 1.4 and 1.5 reproduced by courtesy of the Defense Mapping Agency, Hydrographic/Topographic Center; US Naval Oceanographic Office (2.2); Polar Research Laboratory (2.6); R. Legeckis, NESS (2.11, 5.1); R. E. Cheney (2.12); Figures 3.10, 3.14 and 3.15 reprinted with permission from 'A survey of ocean current theory', *Deep Sea Research* **4** (H. Stommel), © 1957 Pergamon Press Ltd, and by courtesy of H. Stommel; National Weather Service (5.2); Figure 5.3 reproduced from *The Gulf Stream* (H. Stommel 1958) by courtesy of Cambridge University Press; Figure 5.5 reproduced from H. Stommel, *Trans Am. Geophys. Union* **29**, 202–6, © 1958 American Geophysical Union; W. Munk (5.6); G. Veronis (5.7 & 13); N. P. Fofonoff (5.10); the Editor, *Tellus* (5.11); Pergamon Press (5.12); American Meteorological Society (5.15 & 16); M. D. Cox (5.17); R. Heinmiller, US POLYMODE Executive Office (6.6); A. R. Robinson (6.7, 9 & 10); US POLYMODE Organizing Committee (6.8, and for text quoted from 'Dynamics of ocean circulation and currents: results of POLYMODE and related investigations'); F. T. Banner (7.8); L. V. Worthington (7.9 & 13); Figure 7.10 reprinted with permission from 'The abyssal circulation', *Deep-Sea Research* **8**, 80–2 (H. Stommel), © 1958 Pergamon Press Ltd, and by courtesy of H. Stommel; T. D. Foster and E. C. Carmack (7.12); J. L. Reid, W. D. Nowlin and W. C. Patzert (7.14); B. A. Warren (7.15); Figure 7.16 reproduced from *Scientific exploration of the South Pacific*, National Academy Press, Washington, DC, 1970: P. Welander (8.2 & 3); J. Namies (8.5); CLIMAP Project members, *Science*, 1976 (8.6–9).

Contents

List of tables

1 Sea–air interaction

1.1 Interdependence of natural phenomena

Planet Earth owes its unique appearance to processes that occur in three dynamic realms – the atmosphere, the hydrosphere and the biosphere. Ultimately, all the processes within these realms are powered by the Sun. Its radiant energy drives the transport of heat, water, gases and other substances, and sets in motion mechanical processes and chemical and biological transformations. Some of these rhythmic cycles and processes are unchanging; others invariably occur under certain conditions. By their combined actions they ensure the existence of all living things and create the diverse climates and landscapes on the globe as we know them.

Even a cursory glance at the chain of transformations in the three mobile realms reveals an enormous number of natural mechanisms on different scales, a highly complicated pattern of coupled motions, thermal contrasts and interactions between inanimate and living matter. The picture of nature seems so complicated as to defy description. Indeed, one could easily fill hundreds of pages with lists, or even classifications, of observations of the behavior of these natural environments and still fail to provide a clear understanding of the mechanisms, even if there were people patient enough to read such a tedious catalog.

Fortunately, modern science has a powerful tool that enables it to define diverse laws at work in nature and society, in **systems analysis**. What was perceived in the past as a chaos of facts, processes and events can now be organized into an ordered system. From this viewpoint, the Universe appears as a complex system composed of simpler systems of lower rank; each of these, in turn, consists of simpler subsystems, and so on.

The predator–prey relationship, the interlinked evolution of social and economic activities, the interaction between the world economy and the natural environment, and the development of biological communities are all examples of systems where scientists proceed from description and analysis to synthesis and prediction, and eventually to a semblance of control. An artist sees the Universe as a majestic symphony: the joy of this magnificent harmony is revealed, for instance by Churlionis in paintings and music.† A scientist sees the world as a unified system in which all phenomena are linked together in a harmonious pattern at different levels. His goal is to describe these events in terms of logic and mathematics, and to understand how the system functions and interacts with other systems.

The three dynamic realms – the atmosphere, the hydrosphere and the biosphere – form a complex multicomponent system. Its major feature is interpenetration of the individual elements or units of the system. Thus air or its component gases are not just present in the atmosphere, but are dissolved in oceans, rivers and lakes, are contained in rocks in virtually all the Earth's layers, and are active ingredients of organic matter.

Water is as active and ubiquitous as air. The ocean is its main storage reservoir, but huge masses of water are found deep within the Earth (juvenile water), in upper atmospheric strata, in mountain glaciers and in polar ice caps. Like air, water is crucial for life since it is the principal component of every living cell.

The biosphere has no 'headquarters' – it is everywhere. For convenience, the bottom boundary of life is assumed to lie at that surface within the Earth below which the temperature exceeds 100°C. Depending on the structure of the Earth's crust, this surface occurs at depths varying from 500 to 1500 m. The upper boundary of the biosphere lies at 7000–8000 m or more above sea level. All the oceans, rivers and lakes are, therefore, within the biosphere.

The system of the three realms of the Earth, its behavior and evolution are too complicated to portray in any single description. The merit of the systems approach is precisely that it permits the

† Mikalius Konstantinas Churlionis (1875–1911), a brilliant Lithuanian painter and composer, is credited for the proverbial utterance: 'I perceive the Universe as a great symphony, the people as musical sounds. . . .'

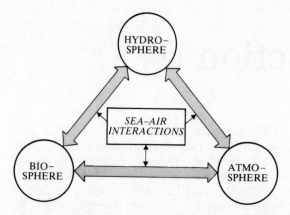

Figure 1.1 Schematic representation of sea–air interaction as a principal mechanism controlling the coupling between the atmosphere, the hydrosphere and the biosphere.

possibility of disaggregating the system, singling out that particular subsystem which carries the main energy burden and thus controls the behavior of the system with the highest rank. The internal regulation of the three dynamic realms is controlled by the perpetual interactions between the global air and water masses, i.e. by the **ocean–atmosphere system** (Fig. 1.1).

The principles of systems analysis require that we define the properties that unite the ocean and the atmosphere in one system. First of all, both play a role in absorption, preservation and distribution of the incoming solar energy. Every minute, the sun delivers 1.94 cal to each square centimeter of the Earth's surface normal to its rays. (The rate of solar heat influx is thus 1.94 cal cm^{-2} min^{-1}, a number usually referred to as the **solar constant**.)

Owing to its high thermal capacity and rapid vertical mixing, sea water absorbs and retains far more heat than does the land. Later, we shall see how heat from the land is important in driving the ocean–atmosphere system through the agencies of evaporation and convection.

Another source of heat is the Earth's interior, which is constantly hot. Recently, deep-sea submersibles have discovered zones of new lava flows and hot springs (the best known are the thermal vents of the Galapagos Rift first studied in 1977). These sources of hot water may locally affect the heat balance. However, the overall internal heat of the Earth conducted through the ocean floor is very small (only 9.0 × 10^{-6} cal cm^{-2} min^{-1}) compared to solar heating, and it is usually disregarded.

The organisms that make up the biosphere also store large amounts of solar energy. Annually, 3 × 10^{21} cal is absorbed by plants through photosynthesis. Of this amount, land plants assimilate a mere 10–20%, the remainder going to marine plants. Although this accounts for a mere 0.04–0.06% of the total solar energy absorbed, it is enormous by human standards. For example, it is equal to the total annual output of 100 000 hydro-electric power stations of the size of the Niagara plant. The energy stored by living things later enters the general energy balance of the ocean–atmosphere system, partly as a result of human activities (combustion of fossil fuel). The energy of tides and volcanoes is also included in the system's general power balance, without ever significantly affecting it. All these 'energy considerations' are cited to bring to the reader's attention that, in its entirety, the natural environment of this planet is largely, if not completely, controlled by sea–air interaction.

The major features that unite the ocean and the atmosphere as one system are their extensive interference and the exchange of internal and mechanical energy and matter between them. Their contact is so intimate that no analytical description of the evolution of either one, during a relatively short period of time, is possible without taking the other into account.

The ocean–atmosphere system can be likened to a machine, and an extremely sophisticated one at that; a machine that has no wheels, gears or wiring. Alternatively, one can compare the ocean and the atmosphere to a cybernetic system in which all processes occur as a result of physical, chemical and biological interactions. Since the most energetic processes involve the transfer of heat and mechanical energy from one medium to the other, the ocean and the atmosphere can be referred to as a **thermodynamic system**. The term 'dynamic system' originated in mechanics, but it is now applied to any material system that requires a transition period of some duration before bringing itself to equilibrium in altered surroundings, rather than adjusting to them instantaneously. We thus arrive at the notion of **inertia** of the interacting media, which will be helpful for an analysis of component subsystems.

The ocean, of course, has a far greater thermal and mechanical inertia than the atmosphere. As we all know, the weather can change overnight, whereas an ocean current may take a month to develop. The timescale of the ocean's thermodynamic response to the seasonal or annual variations in the heat flux is greater, but it, too, is small compared to periods of geologic evolution. For example, the complete cycle of water in the ocean–atmosphere system, defined as the total seawater volume divided by the rate of evaporation from the ocean surface, is completed in a mere 4000 years. Longer cycles in the ocean-atmosphere system, due to cosmic factors and inner system instabilities, are also known. In each case a cycle's duration and regularity depend on the **thermal response** of the ocean. The atmosphere adjusts itself to the ocean rather quickly, and modifies its cycles to fit the new conditions which are determined by the amount of heat stored by the ocean after a particular global cataclysm, or in a particular geologic epoch.

These facts and observations make it possible to divide sea–air interaction arbitrarily and to examine the ocean and the atmosphere as two independent subsystems. This book is largely concerned with the machinery of oceanic movements. They can, however, only be understood and explained if the energy inputs and outputs from the atmosphere to the ocean are given. We shall begin, therefore, with a brief outline of the behavior of the atmosphere, assuming the ocean to be its coupled, underlying environment.

1.2 Cycles in the ocean–atmosphere system

A study of the transformation of energy and matter in the layers enveloping the Earth can conveniently begin with a look at the distribution of heat content among the three different elements – the sea, the air and the land. For simplicity, we will neglect temporarily their geographic characteristics and regard each realm as a uniform subsystem of the ocean–atmosphere system. At this stage it suffices to deal with one or two of the major features of each subsystem, say, heat. We shall begin with the sea.

The distinctive feature of this subsystem has been mentioned earlier – it is the ocean's high thermal inertia, which enables it to store more of the absorbed energy and to retain it for longer periods than is possible for the other two media. The ocean tends to smooth or eliminate, by its internal motions, non-uniformities (a scientist would call them 'disturbances') that might occur

during the transfer of energy and matter among the global subsystems. This 'smoothing function' becomes obvious in the global **heat cycle**. Let us take a closer look at this process.

Given the solar constant, one can readily compute that the upper surface of the Earth's air envelope receives 160 kcal cm^{-2} of solar radiation annually. Of this energy, 70% reaches the surface of the Earth, the remaining 30% being scattered in, and absorbed by, the atmosphere. The solar radiation reaching the Earth's surface totals 122 kcal cm^{-2} a year on average. Part of it is reflected by the surfaces of the sea and land, while another part is re-radiated into space after being absorbed. As a result, the absorbed solar energy, averaged for the planet, is estimated at 70 kcal cm^{-2} a year. This is a little more than one-half the incoming energy.

Owing to the high heat capacity of water and vertical mixing, the ocean can retain 25–30% more heat than the land at the same latitude. In fall and winter, the land loses all the heat it has accumulated during spring and summer. By contrast, the World Ocean has stored, during the long history of the Earth, 500 to 1000 (in some places 1500) times the amount of heat reaching its surface in one year. This is explained by three circumstances.

First, solar energy penetrates to a great depth in sea water. For example, blue-green light penetrates clear sea water to a depth of nearly 40 m with only 50% attenuation. On land, heat does not penetrate deeper than a few centimeters. Secondly, water is one of the substances with the highest heat capacity. Thirdly, ocean waters mix relatively well so that captured heat is distributed to great depths. A 10 m thick later of ocean water carries four times the thermal energy contained in the entire atmosphere, which has an effective thickness of nearly 10 km.

Thus, the picture of heat flux in the ocean–atmosphere system can be schematized as follows. The ocean gets most of its energy directly from the Sun. The atmosphere is supplied by the ocean. The land accumulates no heat over the years, but releases as much heat as it receives, creating extreme seasonal changes of temperature. Thus we see that it is the ocean that must be credited for maintaining the heat content of the Earth at a more or less stable level.

Figure 1.2 illustrates the process of thermal radiation transfer from the ocean into the atmosphere and back. Relatively little energy is consumed on heating the atmosphere through conduction (the so-called sensible heat). This is a

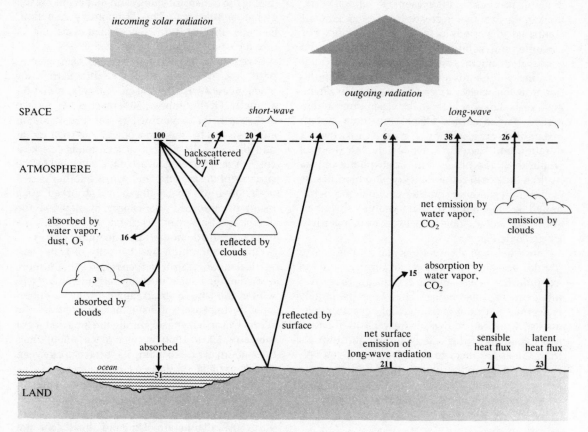

Figure 1.2 The mean annual radiation and heat balance of the atmosphere, relative to 100 units of incoming solar radiation, based on satellite measurements and conventional observations.

mere seven units in the notation of Figure 1.2. Indeed, air molecules cannot capture any large quantity of heat. Atmospheric water vapor has a much greater heat capacity. A large portion of the solar energy is spent on evaporation (latent heat), which can then be released during precipitation.

Most of the inverse heat flux is either absorbed by the atmosphere through evaporation or captured by clouds that block infra-red irradiation into outer space. Combined with carbon dioxide, clouds produce the so-called **greenhouse effect** in the atmosphere. Molecules of H_2O and CO_2 make the lower atmosphere opaque to the Earth's long-wave (infra-red) radiation, while solar radiation easily passes through to the Earth's surface. Heat flux can thus change direction several times and still provide heat to the Earth's surface. The atmosphere would be cooler now than it is, if this 'bouncing' effect did not occur. We shall return to carbon dioxide and the greenhouse effect later.

From the foregoing analysis of the Earth's thermal balance, we have concluded that atmospheric

water has a major part to play in distributing heat over the globe. This brings us to another important atmospheric cycle – that of water, in which the main stabilizing agent is again the ocean.

The **water cycle** starts when water evaporates from all the surfaces found in nature, especially oceans, seas, and continental and island land masses. On average, some 80% of absorbed solar heat is used for evaporation.

As much as 577 000 km³ of moisture leaves the Earth's surface every year. The bulk (505 000 km³) comes from the World Ocean, the remainder (72 000 km³) from the land. Water vapor then condenses and falls as precipitation (Fig. 1.3).

The total annual precipitation over the ocean surface is 458 000 km³, which is less than evaporation. The excess of water vapor, 47 000 km³, is carried by air currents to continental and island land masses, to create rivers, lakes, glaciers and ground water, and provides conditions necessary for maintaining the natural environment and human economic activities. During the year, this

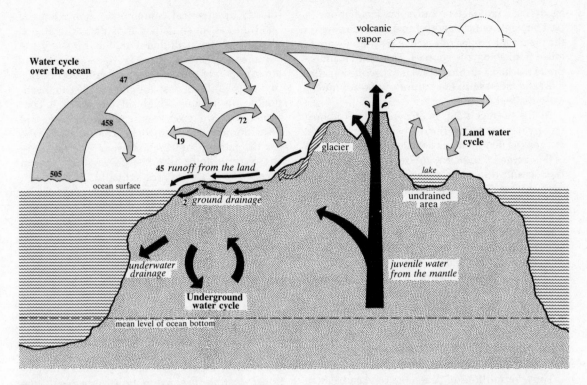

Figure 1.3 Annual water cycle: water transport is given in 10^3 km^3. (Compiled from data presented in Stepanov 1974 and Korzun 1974.)

water returns to the ocean, mostly via rivers (about 45 000 km^3), and partly as ground water not drained by rivers (some 2000 km^3).

Actually, the sea–land water exchange is more complicated. Data from radiosonde observations at high altitude show on-land precipitation (excluding islands) due to ocean water to be equal to 66 000 km^3. The excess of precipitation on land, a total of 19 000 km^3, evaporates once more and is carried to the ocean.

The complexity of the atmospheric water cycle does not end there. An estimate of the total atmospheric moisture at any given moment would be equivalent to a water layer just 2.5 cm thick, that is 1/30th or 1/35th of the amount evaporated during one year. This means that the water vapor content in the atmosphere is renewed completely every 10–15 days. Here we meet with the first important **feedback mechanism**. The excess of water vapor produces cloud systems, which attenuate the influx of solar radiation and reduce evaporation from the ocean surface. The resulting deficit in humidity induces an inverse process. As far as the water cycle is concerned, the ocean and the atmosphere form a self-oscillating system with negative feedback.

Everyday experience suggests that water supply to oceans and continents varies considerably with latitude. Although we have postponed the discussion of geographic variations, it should be mentioned here that the distribution of water to various locations is highly dependent on atmospheric motions and topographic features on the Earth's surface. The schematic water cycle (Fig. 1.3) shows some details such as the intracontinental cycles, undrained areas and the exchange of the Earth's juvenile waters with the surface water bodies.

The foregoing discussion of the Earth's heat and water budgets emphasizes the special role of the ocean in controlling the basic cycles. Its influence on other life-sustaining processes on this planet is no less important. For instance, the ocean helps to maintain a constant gas composition in the atmosphere, especially the proportions of oxygen (O_2) and carbon dioxide (CO_2).

The reserve of free oxygen is maintained by two processes. One is biogenic, in which the oxygen in a water molecule is combined during photosynthesis. The other O_2-releasing process involves decomposition of oxygen compounds at high temperature, by the action of various chemical agents and radioactive and solar irradiation. In past geologic epochs, with life just beginning on the planet, all oxygen released into the atmosphere arose from abiogenic sources. The Earth's atmosphere reached a turning point in its evolution after a drastic increase in the output of photosynthetically derived oxygen.

At the present time, photosynthesis is almost wholly responsible for the renewal of atmospheric oxygen (99.996%). It has been estimated that 78% of this oxygen (i.e. 28.3×10^{12} molecules cm^{-2} s^{-1}) is of oceanic origin and a mere 22% is from terrestrial sources. This is not surprising since, in the sea, microscopic phytoplankton capture solar radiation at depths up to 200 m whereas the tallest trees on land are rarely above 15–20 m tall. Furthermore, the oxygen : nitrogen ratio of air dissolved in water is double that of atmospheric air.

The ocean contains just 1% of the total reserve of free oxygen. One can gauge the efficiency of the ocean as an 'oxygen plant' from the fact that it releases 3.5 times as much oxygen as do all the plants on Earth. Mixing and heating of the upper ocean layers lead to a constant release of oxygen through the sea surface.

Thus, for millions of years, the oxygen balance in the ocean–atmosphere system has been maintained by living organisms. Recently, with the development of giant power industries, humans have begun to interfere with this process. It has been estimated that, toward the end of the century, the burning of fossil fuel will consume an amount of oxygen about equal to the present output by land plants, namely, some $(55–57) \times 10^9$ tonnes per year. A likely result will be a drastically increased amount of carbon dioxide in the atmosphere. The **carbon cycle** in the ocean–atmosphere system is no less important than the oxygen cycle, and deserves a closer look.

The balance of CO_2 gas is affected not only by oxidation processes but also by sedimentation of carbonates and silica in the ocean. Of importance here are biological processes that form inorganic carbonate siliceous skeletons, and photosynthesis, which also takes up some of the carbon dioxide dissolved in water.

The greenhouse effect, due to CO_2 screening of the Earth's long-wave radiation, mentioned above, would become a real hazard in the near future if the amount of carbon dioxide contained in the atmosphere were not controlled by the ocean. This control is possible because the ocean, at equilibrium with the atmosphere, can absorb and hold (mostly in chemical compounds) approximately 100 times as much carbon dioxide as the atmosphere. It is assumed, therefore, that for every 100 volumes of carbon dioxide released into the atmosphere, approximately one volume remains in the air while the rest goes into the ocean. Eventually, the carbon dioxide in the ocean is consumed via two processes – photosynthesis and dissolution of carbonate rocks. The quantity of the gas discharged into the atmosphere by industry and believed to be absorbed mainly by the ocean exceeds 23×10^{11} tonnes annually.

The faster that carbon dioxide enters the ocean, the faster it is absorbed by planktonic organisms. The vital functions of plankton and their death enrich sea water with dissolved organic matter. Its concentration, measured by carbon content at different ocean depths, ranges from 2 to 6 mg l^{-1}.

Consumption of organic matter in the sea by biochemical processes is proportional to the amount available. This helps to maintain its content at a stable level: as soon as the concentration of organic matter exceeds the normal for some reason, the rate of consumption rises accordingly. Conversely, a concentration drop forces a proportional decline in consumption. As a result, the danger of this planet being suffocated by carbon dioxide is greatly reduced.

Here again we observe, in the ocean, a negative feedback mechanism with the initial perturbation triggering processes which eventually level it out. The ocean–atmosphere system seems to be extremely stable and capable of staving off unfavorable developments.

Apart from regulating the exchange of energy and matter on the Earth, the World Ocean exerts a direct influence on the planet's living matter. It functions as an inert medium in the formation of new plant and animal species. To some extent this is confirmed by the distribution of the numbers of land and sea species: only 20% of the existing species are found in the ocean. On land, as well as at the sea/land and sea/air boundaries, living organisms have to adapt to varying conditions, and the rate of evolution there is high. These are environments conducive to accumulation and development of diverse forms of life. By contrast, a uniform environment in the depth of the ocean, and the prevailing low temperatures there, does not promote biological evolution. Besides, mutations induced by cosmic radiation are fewer in the ocean, which attenuates the radiation and thus shields its inhabitants with a dense water screen.

The ocean is not particularly demanding on those of its denizens who have accommodated themselves to the perpetual darkness and chill of its depths. This innate stability explains why relict, 'extinct' animals are found there more often than anywhere else.

The ocean, originally the cradle of life, restrains the evolution of organisms dwelling deep within it. Its inertial power thus harmonizes and 'balances' the evolution of life on the planet, suppressing dangerous evolutionary trends. Even more important for the ocean's regulatory function is that it preserves the genetic stocks accumulated over the entire evolutionary history.

There is no need for us to continue the long list of the ocean's regulatory functions. The foregoing discussion will suffice to provide an idea of its most important general function, which is to smooth out sudden perturbations in the Earth's three realms and to secure a balanced and harmonious course for all the processes on the planet's surface.

1.3 Heat engines in the atmosphere

The foregoing 'compartmentalization' of natural phenomena allows the global exchange of energy and of organic and inorganic matter in the ocean–atmosphere system to be discussed. The above approach was simplified by the fact that, averaged over years, the system's regime is close to **equilibrium**, that is, its inputs and outputs mutually balance each other. However, if one takes a further step and attempts to understand how potential energy in the system is converted into kinetic energy, the above assumption on featureless environments seems fruitless. For this purpose, one must identify the inhomogeneities in the system, caused by heat contrasts between the Poles and the Equator and between the ocean and the continents. These factors either are fairly stable in time or repeat periodically. They are characteristic of the local atmospheric regimes prevailing in the particular regions of the globe, and are normally referred to as **climates**. When speaking of a climate one usually means not only the thermal processes and distribution of water but also the atmospheric motions or winds.

The great naturalist A. von. Humboldt (1769–1859) is credited with the first attempt at a quantitative description of the influence of oceans and continents on climate. His method of mapping geophysical phenomena was adopted universally and has been quite fruitful. One of his earliest maps showed the isolines of surface air temperature. It revealed that climatic zones do not lie parallel to lines of latitude but are affected by the relative positions of sea and land. The existing classification of climates is derived from this discovery.

In addition to warm and cold climates, two more types were distinguished – continental and marine. **Marine climate** is characterized by minor temperature fluctuations and a high humidity. By contrast, **continental climate** exhibits a wide range of temperatures at a moderate humidity. In the heart of Siberia, where the climate is sharply continental, the difference between extreme summer and winter temperatures is 100°C, whereas even in polar latitudes, on the shore of the Arctic Ocean, the difference does not exceed 60°C. In a marine climate, seasonal temperature fluctuations gradually increase from the Equator toward the Poles; in a continental climate, there is almost no such discernible increase.

There are places on the Earth where the wind brings vapor from the sea to the land during the warm season whereas in winter the prevailing winds are offshore. This is called a **monsoon climate**, with a moderately warm, wet summer followed by a cold and very dry winter.

These are simple and evident consequences of the ocean's effect on the atmosphere. V. V. Shuleikin (1953) has schematized this mechanism using a concept of 'heat engines'. An ideal heat engine comprises a boiler, a condenser and the working unit, which utilizes a portion of the energy flowing from the boiler to the condenser. (Incidentally, the size of this portion determines the engine's efficiency.) Shuleikin considered the operation of heat engines of two kinds. In the engine of the first kind (we shall refer to it as the A-engine), the boiler is the Tropics, where the influx of solar energy by far exceeds the back-radiation into outer space. The condenser is the polar regions where more heat is re-radiated than absorbed. To begin with, let us take a simplistic model of the atmosphere's dynamic response to equatorial heating and polar cooling. Imagine that the Earth is stationary and the Sun revolves around it and heats it at the Equator. Moreover, in order to eliminate possible disturbances, let us assume that the surface of this hypothetical planet is made up of a material of uniform heat capacity; for example, that its entire surface is covered with water.

What will the circulation of air be on this hypothetical planet if, initially, the atmosphere is perfectly uniform? As the air warms in the equatorial zone, it will expand and rise. As a result, there will be an excess of air at a certain altitude, and the pressure there will become higher. This is shown in Figure 1.4a. In the meantime, the air cooled near the Poles will contract, and the pressure along the vertical will be distributed in the opposite fashion. The higher pressure belt over the Equator will cause the air aloft to move toward the Poles. Naturally, the loss of air near the Equator will be compensated for in some way, so the wind near the Earth's surface will blow toward the Equator. Thus, on a non-rotating Earth, there will be two cells of circulation situated in the meridional plane (Fig. 1.4).

Since in reality the Earth spins around its axis, this picture has to be modified substantially. The air of an initially uniform planet is first warmed up at the Equator, then rises to a certain altitude and starts its way toward the Poles, but it is immediately deflected by the Coriolis force, to the right in the Northern Hemisphere and to the left in the Southern Hemisphere.†

Therefore, instead of moving as a strictly meridional current toward the Poles, the wind at a certain altitude in both hemispheres will be deflected eastwards and, at about 30°N and 30°S latitudes, will blow parallel to the high-pressure belt over the Equator (as a meteorologist would say, it will blow in the **zonal direction**). The air that accumulates there will descend, forming two belts of higher pressure near the ground (Fig. 1.4b), the so-called **subtropical high-pressure belts**, one in each hemisphere. The sinking air spreads partly toward the Equator and partly toward the poles. Again, the Coriolis force deflects it from the straight meridional direction, forming easterlies (Fig. 1.4b), or winds that close up the two cells of tropical circulation (known as the **Hadley cells**). These Hadley cells are shown around the Equator in the vertical section on Figure 1.4b.

Outside the subtropical zone, two more circulation cells can be observed in each hemisphere. Cooled, denser air from high-pressure polar zones moves toward the Equator. The Coriolis force deflects it westwards, causing easterly winds in the subpolar region. Following the same reasoning, one must expect air to accumulate and rise at

† A detailed explanation of the Coriolis force and how it affects motions on the Earth will be given in Chapter 3.

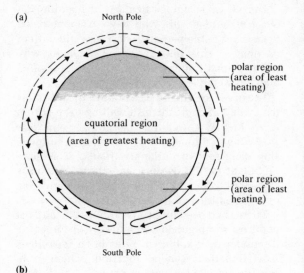

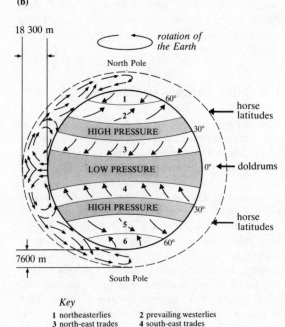

Key

1 northeasterlies	2 prevailing westerlies
3 north-east trades	4 south-east trades
5 prevailing westerlies	6 southeasterlies

Figure 1.4 (a) Ideal wind system on an imaginary uniform, non-rotating Earth. (b) Effect of the Coriolis force on the general circulation of the atmosphere. Two Hadley cells are shown on the left in the subtropical regions. For further explanation see text.

about the 60° latitude in each hemisphere. The intermediate circulation cells between the 60° and 30° latitudes arise inevitably as the links coupling the Hadley cells to the polar cells. Thus, each hemisphere has a three-cell structure in the meridional plane, shown on Figure 1.4b. It can be readily

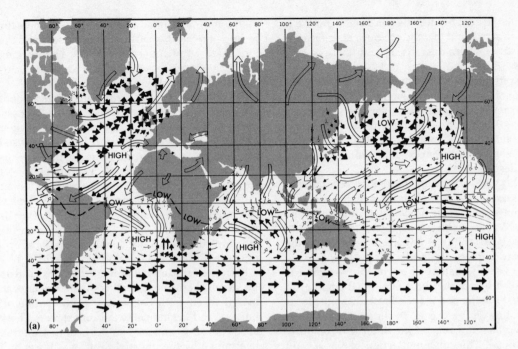

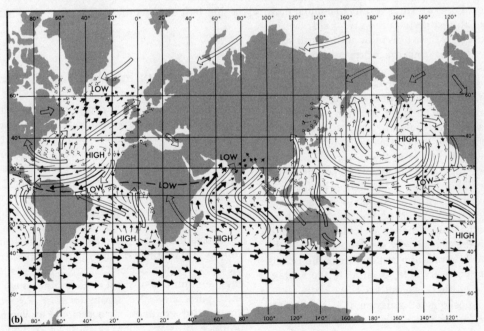

Prevailing winds

Length of arrow indicates generalized
degree of constancy of wind direction;
width of arrow indicates force of wind

→ >20 knots
→ 15–20 knots
→ 10–15 knots
▷ <10 knots
⇨ direction of movement
of air mass

Figure 1.5 Generalized pattern of actual surface winds
(a) in January and February and (b) in July and August.

seen that, at all latitudes, the wind at the ground is close to zonal, and that the pressure distribution† consists of alternating high- and low-pressure zones.

Despite our simplifications, the operation of the engine of the first kind (A-engine) reflects reality. For example, the easterlies over the subtropical regions ('trade winds') and the prevailing westerlies around the 40° latitude (the 'Roaring Forties') are obvious consequences of the natural mechanisms described above.

In this book, describing the atmospheric circulation is not an end in itself. It is just a tool that may help to explain other processes occurring in the ocean. A reader who may be interested in circulation details for the A-engine is referred to the relevant chapters of Strahler's (1971) book.

The zonal transport of air in the A-engine is disturbed by another heat engine (B-engine), in which the boiler and the condenser change places depending on the season: in summer, the land is warmer than the sea and in winter it is the other way round. The combined action of the A- and B-engines of the two kinds results in a much more complex atmospheric pressure field, which determines how the winds blow at the Earth's surface (Fig. 1.5). During the warm season, lower-pressure areas (**cyclones**) prevail over land, and high-pressure areas (**anticyclones**) over sea. Conversely in winter, anticyclones dominate over land (for example, the mighty Siberian anticyclone encompasses the whole of Asia), while cyclonic formations attain great strength over the ocean. However, even in winter the anticyclones of the subtropical regions of the oceans do not disappear completely. For example, the Azores anticyclone over the North Atlantic shifts to the south in winter, leaving the larger part of the ocean subject to the southern edge of the vast Icelandic cyclone. Owing to these relatively stable locations of the high- and low-pressure regions over the ocean, oceanographers are able to estimate the average wind field at the ocean surface (Fig. 1.5).

For the subsequent discussion, it is important to recall that large-scale winds produce the mechanical force that acts on the surface of the ocean and brings it into motion. This is the major mechanical input from the 'atmosphere' subsystem into the 'ocean' subsystem.

The functioning of the two engines is affected by

a third one (designated as the C-engine), where the boiler is the Earth's surface (land or sea) and the condenser is the upper atmosphere. The working unit of the C-engine is the vertical convection currents which give rise to massive clouds.

It is water vapor that possesses the large amount of latent heat that is necessary in the process. When formed at the ocean surface water vapor absorbs latent heat, to release it later (condensed into clouds) to the surrounding air; the air thus becomes lighter and ascends. The pressure near the ground decreases. At the sea surface the air currents move toward the low-pressure center. The wind grows stronger, the mixing of air is increased, evaporation is intensified, and the entire process accelerates. In this way tropical cyclones are formed and develop into hurricanes.

To sum up, the influence exerted by the ocean on the 'atmosphere subsystem' is due to the uneven heating of air from below, and to the supply of moisture and latent heat to the atmosphere. These effects are superimposed on the heat distribution determined by the unequal heating at lower and higher latitudes, which will be described below.

1.4 Sources of instability in the ocean

If atmosphere and ocean conditions were wholly determined by the three heat engines, the weather and climate on this planet would be totally different. All processes would have evolved in a more uniform, simple and periodic fashion. It would have been easier to forecast them, of course (provided that biological evolution under such conditions had produced creatures capable of doing so).

Unfortunately (or perhaps luckily), the above simple heat engines do not operate smoothly, as they have been described. There is a host of heat engines simultaneously at work in the atmosphere and hydrosphere. Numerous local or large-scale temperature gradients, or occasional movements, produce constant changes that we perceive as weather or climatic variation. It is worth outlining the most important sources of instability that affect the entire system.

The ocean's huge thermal and mechanical inertia is responsible for the absorption or filtering out of many small-scale processes, which reveal only decaying traces in the oceanic environment and affect major energy fluxes in the system slightly.

---

† Later, 'field' will be frequently used instead of 'distribution'.

Hence, a time or spatial scale is a major criterion for the importance of various phenomena in the overall energy budget.

Among the three heat engines discussed above, the C-engine has apparently the most limited scale. Actually, this mechanism does not exhibit a regular character – it is largely transient. The powerful thermal convection and tropical storms usually form in a relatively narrow equatorial zone close to the coast, where horizontal thermal gradients are rather pronounced. The hurricanes pass swiftly along the peripheries of larger barometric systems and lose energy over a period of several days. Though the energy of these circulatory patterns is large, it is localized in a limited area and, obviously, does not control the major exhange of heat, moisture and momentum in the entire system.

Owing to huge momentum, the ocean responds to changes in average wind patterns, two of which are shown in Figure 1.5. Therefore, one can assume that seasonal changes in the wind field are one of the large-scale external factors that affect the oceanic movements on the same scale, and thus control the operation of the entire ocean–atmosphere system.

Another source of large-scale instability is associated with heat and water exchange through the ocean surface. Earlier we defined the major processes that convert solar energy into heat and assessed their importance for the entire globe. Now we shall look at their variations with geographic latitude.

Three major components of the heat balance, averaged over each latitude, are shown in Figure 1.6. One sees immediately that two major factors (heat gain by incident radiation and heat loss by evaporation) vary widely in the meridional direction. Since the solar heating at each latitude changes seasonally, the ensuing changes in ocean temperature and ice conditions may affect the A-engine, setting the entire system into oscillatory motion.

The effect of water transport through the ocean surface via evaporation and precipitation is no less important. Clearly, an excess or deficit of water (evaporation minus precipitation) should affect the salt content of the surface layer. Figure 1.7 displays good agreement between the evaporation–precipitation differences $(E - P)$ and sur-

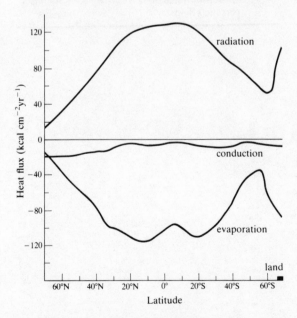

Figure 1.6 Heat transfer to the ocean by each of three major processes as a function of latitude. Positive values mean that the ocean gains heat. (Compiled from data presented in Stepanov 1974 and Korzun 1974.)

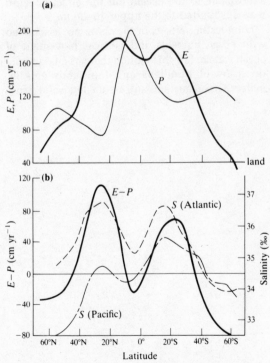

Figure 1.7 (a) Evaporation E and precipitation P as functions of latitude. (b) The difference $E–P$ and surface salinity S in the Atlantic and Pacific Oceans plotted as functions of latitude. (Compiled from data presented in Stepanov 1974 and Korzun 1974.)

face salinity (S) for the Atlantic and Pacific Oceans.

The wind patterns and global heat exchange through the ocean surface are the sources of oceanic processes. It is important to know which oceanic mechanisms can produce or amplify the imposed instabilities that affect the atmospheric processes. In order to identify the major 'carriers' of the external disturbances in the ocean, let us look at the list of the oceanic processes classified in Table 1.1 according to their timescales.

Wave-like phenomena in the open ocean (acoustic waves, tsunamis, internal waves and tidal waves) can hardly affect water structure and induce gradients that cause directional movements. These effects are free oscillations on various scales around some equilibrium position. No matter how large or small these oscillations are, they leave no traces in the environment.

Wind waves have a completely different nature: they are forced oscillations induced on the ocean's surface in response to wind pressure. They are an important mechanism of energy transfer from the atmosphere to the ocean, but the effect of wind waves is limited to the upper 10–20 m.

Long-lasting effects in the ocean are associated with ocean currents and vertical movement of ocean waters. It is this motion that has characteristic scales of hundreds and thousands of kilometers. Ocean currents and associated water struc-ture are responsible for climatic changes; they react to seasonal variations in external heat and give rise to various energy-dissipating mechanisms, such as horizontal eddies of various scales and turbulence.

Ocean currents are the vehicles that carry heat over the entire planet and supply its atmosphere with water, oxygen and salts. They enable the World Ocean to act as a damper and mould the face of the Earth. It will not be surprising, therefore, that they are the subject with which we shall be mainly concerned in the rest of this book.

Table 1.1 Oceanic phenomena and their characteristic timescales.

micro-scales	acoustic waves, turbulence	fractions of seconds
	surface waves, turbulence	seconds
	internal waves, tsunamis	minutes
	surface layer convection, turbulence	
meso-scales	internal waves, tides, deep convection	hours
	small eddies, tides	days
macro-scales	large eddies, synoptic variations in currents	weeks
	seasonal effects	months
	climatic effects	years

2 How ocean currents are studied

2.1 Unity of ocean waters

How infinite and boundless the ocean must have seemed to the first man to set foot upon its shore. Kind or stern, shallow or steep, the ocean's shores have always held a peculiar fascination for man. The moist warm breath of wide expanses of water forces the winter cold to retreat and, in coastal waters, teeming with life, one can always find sustenance. The ever-changing beauty of the sea shore has helped to enrich man's spiritual life. Ancient people could only ponder upon the mystery of existence as they watched inexplicable oceanic phenomena that at times could be very hostile and threatening. How was it possible for capes and distant shores to withstand this great surge, these huge storm waves and tornadoes sweeping across the sky? Only a boundless element could give rise to such phenomena.

In the *Bible*, in the odes of Horace, and in the ancient myths of the Icelandic sagas we find conjectures as to the nature of the ocean – boundless and eternal. In Genesis the ocean existed before the Creation: 'In the beginning God created the heavens and the earth. The earth was a formless void, there was darkness over the deep, and God's spirit hovered over the water' (Gen. 1: 1, 2). The Norwegian skald, Braggi Boddason, describes the way the bulls of the Land of the Giants dragged with huge plows until the island of Iceland was built. In the legends of the people of Oceania, America, China and other lands, the Earth appeared from the ever-existing ocean.

Ancient peoples defied the ocean. Neptune, the Roman god of the sea, and the stern Odin of the Vikings were both abounding in generosity and terrible in wrath. They personified the ocean, the source of gracious mild climate and provider of food, which also terrified man by its powers of destruction, inspiring him with awe for the almighty Nature. In the Psalms of David we read: '... Others, taking ship and going to sea, were plying their business across the ocean; they too saw what Yahweh could do, what marvels on the deep' (Ps. 107: 23, 24).

And when people began 'plying their business' on ocean waters in the quest for wealth and knowledge, the ancient idea that the ocean had no limits eventually proved to be true. Sailing for centuries near the coast, people could still breed hope that somewhere in the west the wide expanse of water would be closed by land. But America, as discovered by Columbus, turned out to be simply a partition separating one ocean from another.

After the conquistador, Vasco Nuñez de Bilbao, saw the Pacific Ocean, previously unknown to Europeans, and the navigator, Magellan, crossed it, it became clear that the seas never end but embrace the entire planet. Moreover, due to Columbus, Vasco da Gama and particularly Magellan, the real proportions of land and sea on the Earth became known. It was learned that continents were simply islands rising out of the water. The great geographic discoveries also made it clear that the ocean determines the conditions on the planet, affecting nature in all her parts. Ever since that time, the science of the seas and oceans – oceanography – has held an important position among geographic disciplines.

Knowledge of the physical properties and movements of the ocean waters was accumulated piecemeal, but it was not until the last century that another, even more striking, feature was discovered, attesting that all oceans are indeed one. It was found that the chemical composition of sea water is the same everywhere. The first round-the-world cruise of the 'Challenger' (1872–6) confirmed this amazing fact. Although overall salt content may vary from place to place, the ratio between individual anions and cations was found to be constant everywhere from the Arctic to the Antarctic. So the concentration of salts in a seawater sample taken from any depth in the ocean is routinely estimated by the content of just one ion, chlorine. This is convenient because most sea-water salts are chlorides (88.64%), the other salts being sulfates (10.80%) and carbonates (0.34%). The ions are in reverse order to the average composition of river waters discharged into the ocean (chlorides 5.2%, sulfates 9.9% and carbonates 60.1%).

When the chlorine content is known, the relationship between overall salinity and chlorine is expressed as

$$S = 0.03 + 1.805\,Cl \qquad (2.1)$$

where S is the total amount of salts (grams per kilogram of solution) and Cl is the amount of chlorine ions (grams per kilogram of water). The unit customarily used for brevity is parts per thousand ($^0\!/_{\!00}$). The likely error of calculation by Equation 2.1 is within $0.02^0\!/_{\!00}$.

There are also other facts that prove the astounding homogeneity of sea water throughout the oceans. A diagram used by oceanographers to describe water masses demonstrates this phenomenon conveniently. At a constant pressure, seawater density is known to depend on temperature and salinity. This relationship is expressed graphically by the so-called **T—S diagram** (Fig. 2.1). Salinity observed in the ocean is plotted along the abscissa, and temperature variation along the ordinate. Seawater density ρ varies

from 1000 to 1040 kg m^{-3}. The quantity $\sigma_t = (\rho - 1000)$ kg m^{-3} is normally used to represent these variations more graphically, and this is the unit adopted in Figure 2.1. The hatched area shows the portion of the diagram corresponding to 90% of ocean water volume. Sea water is thus so uniform that the temperature variation in the bulk of the ocean does not exceed 10°C, and salinity variation is below $1^0\!/_{\!00}$. The variations described in Chapter 1 are observed only in the surface layers. The uniformity of the entire water body results from mixing by major ocean currents.

2.2 The major ocean currents

An ocean current is a natural phenomenon on a breathtaking scale. Enormous masses of water are carried by it over thousands of kilometers. In contrast to the steadier course of a river, an ocean current may become wide and then diffuse, change its direction several times and even turn back in its flow. Warm water in the western parts of the oceans flows poleward and, like a water heating system, warms up the higher latitudes of the Earth, is cooled, and returns to the Equator. What amounts of water are involved in this giant turnover? The Gulf Stream alone carries 50–70 times as much water as all the rivers on the Earth.

Suppose a pilot in a spaceship hovers over some point of the ocean, as might a communications satellite. The ocean circulation displayed on the screen of his radiation temperature recorder might remind him of a chart of a giant animal's blood vessels. This picture would be very similar to Solaris, the cosmic ocean from Stanislaw Lem's science fiction novel.† In fact, this plot could only arise in the mind of a dweller on Earth for whom the ocean is the source and foundation of all life. Incidentally, the notion as such is not new: Kepler (1571–1630), who formulated the laws of planetary motion, earnestly considered the possibility that the Earth and other celestial bodies might be living organisms. He was prompted to this idea by ocean floods and ebbs, which he interpreted as the planet's breathing.

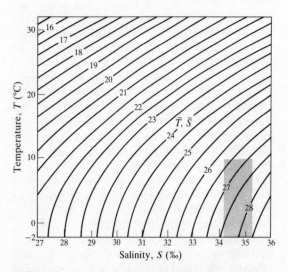

Figure 2.1 *T–S* diagram is a graphic representation of the relationship between density σ_t at atmospheric pressure, temperature *T* (°C) and salinity *S* (‰); 90% of the ocean volume has temperature and salinity ranges within the hatched rectangle.

† Stanislaw Lem, a contemporary Polish writer, is known for his brilliant science fiction. In his novel *Solaris* (1978), he describes a planet which is a living organism, resembling ocean in form but possessing an enormous mind with the faculty of recreating visual and tactile images.

The first thing to note in a panoramic view of ocean currents is that despite the various shore and bottom geometries of the three major oceans – Atlantic, Pacific and Indian† – they all exhibit a nearly identical pattern of circulation. Let us take a look at the schematic diagram of surface currents in the World Ocean (Fig. 2.2.)

We begin with the closed rotating systems of currents, usually called **gyres**. Some of them have the shape of a narrow elongated ellipse, but rotation is observed in these as well. Two directions are usually distinguished: **meridional** (along the meridian) and **zonal** (along the parallel circles). The clockwise rotation is called **anticyclonic**, and the counterclockwise rotation **cyclonic**.

First, let us look at the large subtropical gyres, the first circuit in the Earth's 'heating system', found in all three oceans. The rotation is anticyclonic. The low-latitude parts of a gyre are formed by the Northern and Southern Equatorial Currents. These are wide (up to 2000 km along the meridian) and stable currents with an average velocity of 20–50 cm s^{-1} (but sometimes as high as 100 cm s^{-1}). In all the oceans, the currents have an east–west orientation in the zone between the Equator and the Tropic.

The Gulf Stream and Kuroshio, the generally known geographic names, are the western components of subtropical gyres in the Northern Hemisphere – rather narrow (100–500 km) fast streams flowing close to the western boundaries of the oceans (hence the other name of these flows – boundary currents). At some moderate latitudes both currents leave the coast. Some streams branch off from the main flow and form equatorward currents. However, the bulk of the Gulf Stream and Kuroshio waters flows eastwards and later forms the North Atlantic and North Pacific Currents, respectively. These, in turn, partly feed the gyres at northern latitudes (most intensively in the Atlantic) and give rise to unstable southeasterly currents (the Portugal and Canary Currents in the Atlantic, and the California Current in the Pacific), closing the huge subtropical gyres in the Northern Hemisphere.

In the Southern Hemisphere, the subtropical gyres are even more impressive, although their western parts (the Brazil and East Australia Cur-

rents) are weaker than their northern counterparts – the Gulf Stream and Kuroshio. In the Pacific, the eastern part of the gyre is the cold Peru Current (also referred to as the Humboldt Current) which has an interesting peculiarity: it is the strongest and narrowest of the currents near the ocean's east coast. In the western part of the gyre in the Indian Ocean, the strong Agulhas Current makes a sharp contrast to the weak diffuse West Australia Current in the ocean's eastern part. Then the Antarctic Circumpolar Current or West Wind Drift is most notable among the currents in the World Ocean. It is a powerful and deep (2500–3000 m) flow with an average speed of 25–30 cm s^{-1} that crosses all three oceans to close up the southern subtropical gyres.

The gyres of the moderate and high latitudes are cyclonic. They are seen distinctly in the northern parts of the oceans, especially in the Atlantic. The strong North Atlantic and North Pacific Currents are to a large extent responsible for the intensity of the northern gyres. The currents are complex in the North Atlantic north of 50°N as this area is affected by free water exchange with the North Polar Ocean, and has an extremely complicated bottom relief. A number of well developed gyres can be found, located to the north-west of the North Atlantic Current. In turn, it partially feeds the warm Irminger Current, but the main bulk of its water advances further north in the form of the Norwegian Current, then enters the Arctic Basin. Owing to these warm flows, Iceland, Spitzbergen and northern Scandinavian shores enjoy mild winters, quite unusual for those latitudes. The cold East Greenland Current makes its way from the Arctic Basin along the Greenland coast, closing up all the gyres and eddies of moderate and polar latitudes in the Atlantic. The southernmost continuation of the cold flows, the Labrador Current, penetrates deep into the south, meeting there with the Gulf Stream.

In the Pacific, two distinct gyres are observed: in the Gulf of Alaska and in the Bering Sea. The gyre's western component, the cold Oyashio Current, goes south, cooling all the coasts of the Far East. The North Pacific Current brings far less heat to the higher latitudes than does its North Atlantic counterpart. Unlike in the North Atlantic, the pattern of the continental coastline of the North Pacific forces the warm currents to change course and turn southward.

The subpolar eddies near the Antarctic are feeble and unstable. They only exist in the Ross, Bel-

---

† The North Polar Ocean is so different in size and water regimen from the other three oceans that it is frequently regarded as a sea of the Atlantic Ocean – the Arctic Sea or Arctic Basin.

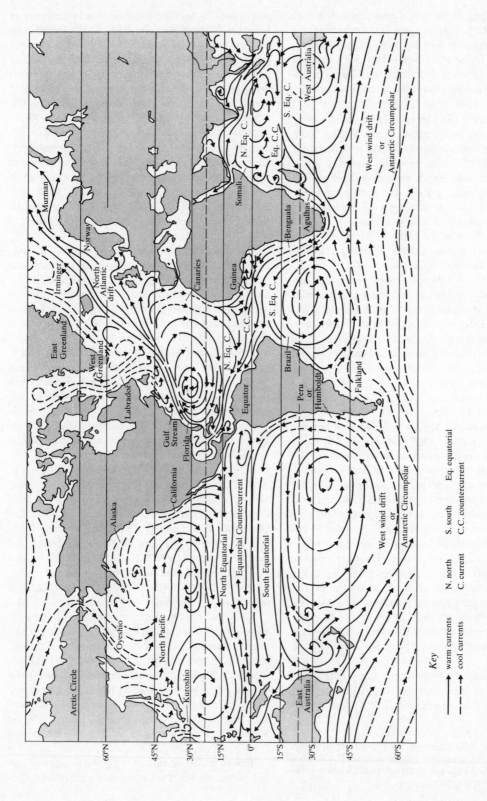

Figure 2.2 A map of the major surface currents in the World Ocean (winter in the Northern Hemisphere).

Key

→ warm currents   N. north   S. south   Eq. equatorial

--→ cool currents   C. current   C.C. countercurrent

lingshausen and Weddell Seas, cutting deep into the land. The southern component, the Antarctic Polar Current, is unstable and slow – 2–3 cm s^{-1} (rarely 25 cm s^{-1}).

Current measurements in the Arctic Basin have been obtained fairly recently, mainly by Soviet polar expeditions. There exists just one anticyclonic gyre. However, its structure is disrupted by local circulations in the seas, and by the polar flow originating from the Chukchi Sea and passing over the North Pole to Greenland. The currents are slow (1–3 cm s^{-1}), growing stronger only in the straits between Greenland and Scandinavia.

The equatorial and tropical zone is believed to be one of the most dynamic parts of the ocean. For the past two decades, it has been the focus of many research oceanographic projects. The reason for this interest is the fact that the system of equatorial currents largely determines oceanic and atmospheric motions at other latitudes. In fact, this is the 'weather kitchen' of the Earth.

It has been known since Columbus' time that the trade winds in the Tropics give rise to the mighty equatorial currents in both hemispheres, divided by a belt of calm seas and weak winds. In this weak-wind zone, called doldrums, the Equatorial Countercurrent is observed, which flows in the opposite direction to its northern and southern neighbors. Every ocean has such a system of currents and countercurrents, although with some local peculiarities. In the Pacific, the countercurrent starts near the Philippines and flows directly eastward slightly to the north of the Equator between two equatorial currents. It bisects the ocean (for about 13 500 km) at an average velocity of 40–60 cm s^{-1}. In the narrower Atlantic Ocean, a portion of the Southern Equatorial Current reaches the Northern Hemisphere and merges with the Northern Equatorial Current. The countercurrent originates 1100–1600 km off the South American coast, flows eastward and, gradually gaining strength, feeds the Southern Equatorial Current.

In the Indian Ocean, the system of equatorial currents is shifted south of the Equator and is strongly affected by monsoon winds (which have been discussed in Ch. 1). During the northern winter (December through February), the time of the Northeast Monsoon, the equatorial system operates 'routinely': as in other oceans, there are equatorial currents and countercurrents. The Somali Current, the analog of the Gulf Stream and Kuroshio, behaves in a 'strange' manner, moving southward in a wide band. In summer (June–August), when the Southwest Monsoon prevails, the Equatorial Countercurrent disappears while the Somali Current becomes a narrow northward flow, even more than the Gulf Stream.

In the equatorial zone, subsurface currents, the Cromwell and Lomonosov Undercurrents, have recently been discovered, although they are closely linked with the system of near-Equator currents. We shall only be concerned at this point with surface currents and return to these deeper currents later.

2.3 General ocean circulation

We have seen that systems of surface currents are astonishingly similar in all the oceans, despite minor differences. Without going into the reasons for this similarity – obviously related to similar governing factors – we shall now take it for granted and try to build a generalized scheme of currents in a hypothetical 'typical' ocean, sometimes referred to as **horizontal ocean circulation**.

Subtropical gyres and equatorial countercurrents are the principal elements of the circulation. The eddies at moderate latitudes are less stable and smaller in size. Schematically, these features of currents were demonstrated for the first time by O. Krummel (1911) and later by N. N. Zubov (1947). Both represented the ocean by a simple figure: a rhombus with rounded angles, or an ellipse (Fig. 2.3). Inside the figure, the two principal gyres are shown to be symmetrical in relation to the Equator. In Zubov's presentation, the gyre of the moderate latitudes is indicated by small arrows. In these generalized diagrams of ocean currents, the specific features of each hemisphere are disregarded.

V. N. Stepanov (1960) suggested a more detailed diagram of the ocean circulation. He scaled down the zonal sizes of each ocean proportionally (except for the Arctic Basin). Between 60° and 70°S no shore is shown since, except for the island arc between Tierra del Fuego and Antarctica, the ocean is virtually continuous in this zone. The diagram includes all major gyres and currents of the World Ocean. It shows, for example, that the northern subtropical gyre, as compared to the southern gyre, is shifted westward, and that the western currents in the north (the Gulf Stream and Kuroshio) are narrower and faster than their southern counterparts. Details such as the deflection of the northern component of the sub-

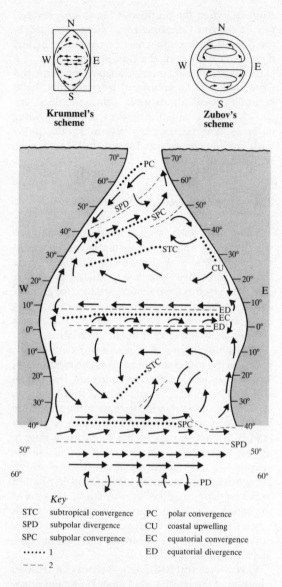

Krummel's scheme   Zubov's scheme

Key

STC	subtropical convergence	PC	polar convergence
SPD	subpolar divergence	CU	coastal upwelling
SPC	subpolar convergence	EC	equatorial convergence
······ 1		ED	equatorial divergence
– – – 2			

Figure 2.3 The general circulation of surface waters as depicted by Stepanov (1960). Krummel's and Zubov's schemes show some earlier representations. 1 – lines of convergence, 2 – lines of divergence.

The diagrams of ocean flows (such as Fig. 2.3) are also helpful because they provide a general picture of currents. It would be difficult to illustrate a theoretic interpretation of the currents, considering real currents and gyres in all their complex detail. A theoretician always prefers to deal with an abstract entity that represents the main features of the real object. This function is well served by the mnemonic diagram in Figure 2.3, which will help the reader to distinguish between the subtropical and moderate-latitude gyres.

In a sense, the diagram also illustrates the general principle of studying ocean currents. Since it is so easy to reduce the actual circulation patterns in all three oceans to a single abstract diagram, it may be reasonably assumed that general governing factors could be found behind this unity. Since gyres are the basic units of the picture, it will be sufficient at the start to describe the mechanism of just one major gyre. Conclusions derived in this way will then be substantiated by illustrations drawn from the favorite testing ground region of oceanographers, the North Atlantic, the most developed and explored part of the World Ocean. After the general features, other more subtle dynamic phenomena, such as the westward shift of the gyres, or the formation of countercurrents and eddies, will be described.

In conclusion, one shortcoming of the circulation diagram should be pointed out: it represents motions in a plane, showing only the surface currents, while the real oceanic circulation must have a complex vertical structure as well. In order to provide at least some indication of the vertical motions, most diagrams also show **convergence zones** of water masses of different origins, where waters usually sink, and **divergence zones**, where currents move apart and water rises toward the surface. The convergence line is called a **front**.

2.4 Exploration of ocean currents and their use in practice

Knowledge of currents is obviously valuable in coastal zones where human activities such as fishing, sailing, shipbuilding, construction of harbors and provision of recreation facilities are carried out on a large scale. Charts and manuals are used to help forecast the speed and direction of tidal currents in harbors, channels and fairways, and are systematically updated. Currents are also investigated at sites of shore reinforcement, con-

tropical gyre from the exact zonal direction, the penetration of cold northern currents southward along the west coast, and the absence of meridional transport below 40°S are depicted in Figure 2.3 by the surface current systems which are developed by the combined action of the external factors and the peculiarities of the coastline. The diagram therefore calls for a theoretic explanation of the patterns of surface water movement in each particular case.

struction of ports and channels, discharge of industrial and municipal waste, etc. The practical value of knowing currents in the open sea is not as evident. Seamen were obviously the first to notice their existence. There is evidence that peoples of the ancient Mediterranean civilizations had some knowledge of currents in the Atlantic. Ancient seafarers – the Phoenicians and their descendants, the Carthaginians – ventured out of the Mediterranean to trade with the inhabitants of the Canaries. It is probably from them that the ancient Greeks acquired the notion of the ocean as an immense wide river flowing along the side of a flat continent. After Aristotle, who conceived the Earth as a sphere, the notion of a river gave way to that of a basin where the water flow has no prevailing direction. A flow reminiscent of a river could not be observed in the Mediterranean except in straits, and Aristotle, and his pupil Theophrastus, mentioned currents in the Straits of Kerch-Enicale, Bosphorus, Dardanelles and Gibraltar.

In the Middle Ages, only the Norsemen continued to explore the maritime routes, sailing first to Iceland and later to Greenland and North America. Their knowledge of movements in the ocean is reflected in the names they gave to promontories and islands, such as Island of Currents, Currents Cape, and so on.

During the epoch of great geographic discoveries, much was learned about currents. First the Genoans, and then the Portuguese, became familiar with currents near the African coast, particularly the Guinea and Benguela Currents. Late in the 15th century, Vasco da Gama, on his voyage to India, noticed the Mozambique Current.

A historian of the exploration of currents is faced with great difficulties because, from the time of the Phoenicians till long after Columbus, sea charts were held in secret by navigators. Spanish conquistadors had their charts attached to lead plates which they threw into the sea when the ship was attacked by pirates. However, it is known that Columbus was the first to use instruments to determine the direction of the current flowing north of the Equator. Sinking the sounding lead from his caravel and watching the deflection of the line, he found that everywhere along the route to the West Indies the currents ran opposite to the rotation of the Earth, and thus he discovered the Northern Equatorial Current. The records of Antonio de Alaminas, the navigator of the Spanish conquistador, Ponce de Leon, show that in 1513 he discovered the Gulf Stream. Ironically this expedi-

tion, launched in search of a 'river of life' by an aging fierce conqueror who hoped to be healed of the ailments of his old age, indeed resulted in the discovery of the oceanic 'river' that supplies heat and moisture to the whole of Europe.

A great amount of historical data on currents collected by fishing skippers has certainly been lost: these data only became common knowledge after scientific analysis was introduced into their collection. Back in the Middle Ages, intelligent people understood the value of seamen's observations, and even of superstitions and legends that mentioned information about currents. From the mid-17th century onward, maps of currents began to appear in the works of naturalists.

The first maps, compiled by Fournier in 1643, Varenius in 1650 and A. Kircher in 1678, still contained a good deal of fantasy. In 1786 however, a map was published that was perfectly suited to practical use at that time – its author was Benjamin Franklin. This map showed only one current, the Gulf Stream, with remarkable accuracy, and was compiled expressly for one practical purpose. Captains of English packet boats had been notoriously late with the delivery of mail from England to the New World. They would not listen to American whaling captains who recommended avoiding strong countercurrents near the American shore. With characteristic assiduity, Franklin, at that time Postmaster-General of the British Colonies, studied all the information on the Gulf Stream collected by whalers and fishermen, traced the current over most of the ocean and drew its first map.

In the 19th century, routine collection of data on currents from navigators was organized due to the efforts of Alexander von Humboldt, James Rennel and, especially, Matthew Maury. The chronometer, invented in 1767, and the sextant enabled seamen to determine their position in the ocean with good accuracy, and to calculate their ship's drift off course due to currents and wind. From their data, Maury compiled maps of ocean currents. In 1835, the First International Meteorological Conference was held in Brussels, and it was decided to set up a uniform system of marine observations. From that date onward, information has been collected systematically on currents, wind, water and air temperatures, atmospheric pressure and precipitation, leading to the compilation of maps and navigation guides.

A thorough exploration of the deep ocean waters started with the four-year British expedition on the

'Challenger' (1872–6). Since then a great number of ships under various flags have put to sea to collect information which has been used to draw charts of ocean circulation.

Throughout the history of oceanic exploration, data on currents have been gathered almost exclusively by and for seamen. With the transition from sail to steam, interest in currents gradually subsided. This loss of interest was furthered by improved methods of navigation and higher ship speeds. However, after the wreck of the 'Titanic' in 1912, concern about collisions with icebergs prompted studies of currents carrying polar ice. Since that time, a special ice patrol service on seafaring routes from Europe to America has been monitoring icebergs. By observing movements of drifting ice, it was possible to study the Labrador Current and the Gulf Stream in greater detail. Yet even today, with advanced navigation methods, the approach to Canadian and US shores is sometimes dangerous because of combined fog and ice hazards.

Data on open-sea currents found in atlases and manuals are rarely taken into account when modern navigators, with perfect radio navigation equipment, map out a route. Efficient routes are recommended on the basis of wind system forecasts, particularly considering the probable storm areas. As to currents, the navigator does not need to introduce in advance any corrections for the ship drifting off course. By taking his bearings at sea frequently and with sufficient accuracy, he can do without preliminary data on currents. Conversely, by investigating the corrections for drift that were actually made by the ship, one can get an idea of the currents it met on its way. Maury used this method to map ocean currents.

At present, the practical applications of currents and other environmental characteristics are numerous. The following description briefly outlines the major uses of information on ocean circulation and properties and their influence on economic activities related to the ocean.

Modern navigation requires knowledge of the actual flows in the oceans. The need for precise information on current fields at various depths has become acute for submarines, as their underwater residence time has increased. Submarines must take into account drift caused by deep currents. The variability of currents has proved to be so wide that, even in areas of persistent winds (as in the tropical ocean – see Section 6.6), the average currents represented on climatic charts in atlases can be misleading for submarine navigators. Short-term forecasts are required. The disaster of the loss of, and subsequent search for, the US submarine 'Treasure' illustrates the importance of knowing the deep currents.

Fishermen also need information about ocean currents, temperature and salinity of the upper ocean, but their requirements are broader. The widest varieties and highest concentrations of plankton occur in zones of temperature–salinity contrasts, which sustain the food chains of larger marine animals. Such areas are found in regions where waters of different origins come together. Earlier (Section 2.3) these areas were referred to as frontal, or convergence, zones. In the mouths of estuaries, where river flows meet sea water, at the boundaries of cold and warm currents, or in places where a current approaches a countercurrent, plankton are usually found in abundance, giving food to innumerable schools of fish which, in turn, attract large numbers of sea birds.

It was found that some species of fish are sensitive to temperature differences as small as 0.03°C. Temperature affects the maturation of the gametes in fish, the incubation time of the eggs, the larvae survival rate and other life functions. Fishery forecasts and successful catches are largely based on prediction of temperature and current fields in fishing areas.

However, the ocean abounds with life, not only at such 'contacts'. An interesting phenomenon of upwelling is observed, for example, near the shores of Chile and southwestern Africa. It occurs in places where the cold currents (Peru, California, Benguela and others) flow close to the coast, and it essentially consists of the following. In surface layers, there is constant water transport into the open sea, separate from the longshore flow. Frequent winds increase this drift of the upper layers away from the shore. As a result, deep waters rich in oxygen and nutrients rise to the surface, creating conditions for a fabulous wealth of species, especially fish. There are some years, however, when the upwelling may diminish because of a change in velocity of the longshore current (see Section 8.1), which leads to the disappearance of plankton and to the resulting mass mortality of fish and other sea creatures that are deprived of their food supplies. Forecasting currents in upwelling areas is therefore extremely important.

In recent years, technological means have been developed for survey, exploration and extraction of mineral resources from the ocean bed. The most valuable ocean resources are oil, gas and sulfur, which are found in many coastal areas. Phosphorite deposits and manganese nodules are potentially usable resources of the offshore areas and deep basins.

Presently, about 20% of the world's petroleum products come from the sea, and this figure may reach 50% by the year 2000. Drilling for gas and oil is now limited to depths less than 200 m. However, exploratory wells have already been drilled to a depth of about 700 m. A very expensive system for the positioning of a drilling rig has been developed, to offset the action of wind and currents and to hold it stationary above the drilling site. The system makes it possible to re-enter a borehole when the worn bit is to be replaced on the drill string. The deep-sea research drilling vessel 'Glomar Challenger' has performed drilling and coring for ocean sediments in all the oceans. This drilling has mostly been for scientific purposes, and has been successful at a depth of over 6000 m.

Inhospitable environmental conditions in the ocean affect marine surveys, exploration and mining operations. Specially equipped ships are required to undertake dredging and extraction of deep bottom sediments, and excavation of sand and gravel in coastal waters. Data on ocean dynamics are also important to the design and installation of pipelines on the sea bed. Tides and storms create currents that act on these pipelines. The safety of these facilities is becoming increasingly important as larger-diameter pipelines are being planned for offshore areas all round the world. Forecasts on currents are vital in icy conditions, and are now extremely important for those involved with offshore drilling in the Arctic regions.

Computer-based numerical forecasting of ocean conditions and weather is based on information about current fields in the open ocean. An insufficiency of this information is a major factor that limits the accuracy of climatological data and ocean weather forecasts.

Studies of ocean currents have been gaining impetus due to attempts to assess the environmental pollution of marine bodies and to impose effective pollution control. In some coastal areas, environmental pollution has reached an appalling level, threatening to impede further economic development. Certain ecological effects, such as eutrophication in some estuaries, oxygen deficiencies and mass mortalities in previously healthy habitats, and spreading of hydrocarbons

and DDT into very remote areas, call into question the ability of the ocean's natural mechanisms to break down some man-made pollutants. The problem seems to be very complex, because very little is known about the present level of pollution compared to 'pristine' conditions, and about new cycles that may evolve to offset the man-induced changes in ocean chemistry and biology.

Some pollutants (such as pesticides, lead and mercury compounds, etc.) enter the ocean virtually uncontrolled via atmospheric circulation. Certain wastes (such as agricultural chemicals, petroleum residues, industrial and municipal effluents) come to the ocean via rivers. These sources are most abundant. Direct ocean outfalls are believed to be under a certain amount of control and may affect only limited areas.

In the early 1970s, the Soviet Union launched a multi-million rouble program to investigate pollution loads and fields of major toxic compounds in the North Atlantic, with emphasis on the Gulf Stream region, the coastal waters around Europe and various basins of the Mediterranean (Simonov 1979). Four weather ships, affiliated to the USSR Hydrometeorological Service, were involved in the work.

The study showed that the major industrial regions in Western Europe and North America were sources of noticeably elevated concentrations of heavy metals, oil products, organochlorine compounds, etc., deposited over vast areas of the ocean. These substances enter the ocean mostly via rivers, the atmosphere, or, perhaps, due to occasional spills and dumpings. The dilution of effluents or initial patches of pollutants appeared to be less effective than was expected. Ocean currents transport pollutants over long distances, partially dumping them into stagnant waters within the anticyclonic gyres (Section 2.2), where their residence time is fairly long. For instance, the eddies in the Sargasso Sea are marked by high concentrations of oil residues. The concentration of mercury in water over the George's Bank was found to be so high that the rate of photosynthesis decreased, as compared to previous years. In land-locked seas, such as the Black Sea, some basins of the Mediterranean and the Baltic Sea, the prevailing counterclockwise current systems keep the center of the seas fairly clean, but the coastal water remains heavily polluted instead.

These and other studies indicated that some marine pollution monitoring is required. A pilot project called the Marine Pollution (Petroleum) Monitoring Pilot Project (MAPMOPP) was conducted in the late 1970s, in a form of a baseline ocean-wide survey to delineate the areas of significant contamination, pollution inputs, pathways and exposure criteria (Levy *et al*. 1981). MAPMOPP has set up the organizational framework for future marine pollution monitoring programs and generated a great deal of information concerning the global distribution of several forms of oil pollution. However, little or no information is available for very remote and unexplored areas. Information is scanty on the transport of oil residues by surface ocean currents from the areas of input. The succeeding project, Marine Pollution Monitoring Program (MARPOLMON), will provide a more detailed and regular surveillance of contamination in the World Ocean.

The importance of information on the dynamic and thermal characteristics of the ocean will increase in future ocean management. Several so-called 'innovative uses' of the ocean (Ross 1978) call for more intense exploration of its energy and food resources.

One long-disputed project to generate hydraulic energy from the difference in water level between the seas, known as the Atlantropa Project, was developed in the 1930s. It proposed that hydroelectricity should be produced in the Strait of Gibraltar by building a dam to section off the Mediterranean Sea from the Atlantic Ocean. The huge evaporation from the Mediterranean would maintain the head of water necessary to operate a giant power plant. Such 'evaporation' plants could be built in other seas as well. Although serious doubts have been voiced about its efficiency, the Atlantropa Project has not been shelved completely. The major objections are that the Gibraltar dam may disrupt the ecologic balance of the Mediterranean, leading to deterioration of this unique sea. The dam would also prevent the constant influx of the dense Mediterranean water into the Atlantic, which would cause presently unknown consequences in deep-sea circulation and, perhaps, would lead to inadvertent climatic modification.

A more realistic and less environmentally damaging idea is that of energy generation based on ocean thermal energy conversion (OTEC). In a tropical ocean a temperature difference of about 15–20°C is observed in a 1000–1500 m deep layer. This temperature gradient is very stable, because it is maintained by global circulation (Section 2.2),

and therefore it could form a promising abundant source of energy for the future. However, two OTEC systems (one designed by TRW and Global Marine, the other by Lockheed) have demonstrated that the large capital investment and high operational costs, with limited upper levels of output, make such power plants relatively inefficient. With the increasing cost of energy in the world, OTEC systems should become more attractive. But even in this case, OTEC plants will be site-specific and will not become a major source of energy.

Other innovative uses of the ocean include attempts to enhance biological productivity by increasing the rate of chemical reactions and mixing processes. Currently, there is a danger of overfishing some important species, which may lead to dire consequences in the ocean's productivity.

One of the ways to keep fish production at a high level is artificial fertilization. Basically, this is the special enrichment of sea water by substances that facilitate rapid development of marine organisms. Of course, this method is applicable only in limited areas of an ocean's coast or, preferably, in a closed water body. The well known Soviet hydrobiologists, Zenkovich and Marti (1970), estimated that 3000 tonnes of phosphates could increase the annual fish output from the northern Caspian Sea by 2000 tonnes. The key to success in this operation is exact knowledge of the sea currents and mixing processes at various depths. It is expected that the fertilizers would be suspended within the water column for a long time so that they could be efficiently used by marine organisms.

Other avenues for oceanic fish breeding are also envisioned. V. S. Latun (1971), who has studied upwelling for many years, believes that a more promising future technique would be to activate the natural processes responsible for enriching the life-supporting sea layers with nutrients. In particular, he suggests artificially intensifying the water rising near shelves to simulate natural upwelling. The idea seems especially attractive since it allows controlled feeding of all organisms – from the tiny algae to the higher forms – by nutrients raised from the depths.

Two other proponents of artificial upwelling, Isaacs and Schmitt (1969), suggested a more realistic scheme. Their project calls for utilization of waste heat from future thermal or nuclear plants to stimulate vertical circulation. Tentative estimates indicate that the introduction of nutrients from deep water into the zone of photosynthesis over a

1.5 km length of the USA coastal zone would be sufficient to produce an additional amount of animal protein to satisfy the requirements of 2000–2500 people.

At this time, however, there are no technical facilities for raising such huge quantities of water to the surface, although some day artificial upwelling may become a tenable economic proposition. The World Ocean will then become the 'global pantry', supplying the planet with various mineral and biological resources.

2.5 Current measurements

Data on currents collected over many years provide a general picture of ocean circulation. However, despite the advent of modern technology, the movements of ocean waters at any given moment are virtually unknown. Unlike the atmosphere, for which environmental observations are performed routinely by a global synoptic weather network, there are no regular reports (with a few notable exceptions) on the conditions of the ocean. The reason for this is not only the inhospitable conditions of the ocean, which greatly harass data accumulation, but that world economic development has not yet reached a level at which it is beneficial to incur enormous costs for the design, construction, deployment and operation of a network of stations for continuous deep-sea monitoring. This section attempts to unify the development of ideas about major observational procedures concerning ocean currents and related phenomena.

Historically, the basic features of ocean currents were obtained using so-called indirect measurements. Owing to high thermal inertia and slow mixing, particularly in the deeper strata, ocean waters preserve the temperature, salinity and other characteristics of their 'origin'. It was known long ago that the warm and cold currents are well manifested in the surrounding waters. Thus, simple mapping of temperatures of the surface waters allowed easy identification of various flows.

It was mentioned in Section 2.3 how in 1786 Benjamin Franklin used data on the marked temperature contrasts observed by fishermen and sailors to map the Gulf Stream. A little later, in 1797, Count Rumford made a remarkable discovery of the deep, cold flows from the polar regions to the Equator, using very few temperature measurements and experiments on convection in liquids (cited in Warren 1981). Later Humboldt in 1845 generalized the idea of using the temperature distributions in the ocean and atmosphere to discover the sources of the major flows.

At the beginning of this century, a rigorous technique was developed to compute the so-called geostrophic currents from temperature–salinity distributions (the method is discussed at length in Section 3.4). These geostrophic currents were found to resemble closely actual large-scale circulatory patterns in the oceans, and temperature–salinity data have become the major source for computation and mapping of the principal features of ocean circulation.

In the 1920s and 1930s, several research cruises were carried out with the primary goal of computing geostrophic currents. These studies were later facilitated by the invention of the bathythermograph (BT) which allowed the rapid recording of vertical temperature profiles. The BT operates as follows: A pressure-driven bellows moves a metal- or smoke-coated glass slide under a stylus by a liquid-filled Bourdon tube sensitive to temperature. With the advent of electronic technology, thermistors were used to sense temperature, and a sophisticated expendable bathythermograph (XBT) was invented for nearly instantaneous temperature profiling to a depth of 1830 m. A new conductivity(salinity)–temperature–depth (CTD, STD) system has replaced very slow chemical determinations of salinity. However, even these methods appear to be not fast enough to track rapid changes in the surface currents. Therefore, space technology has been used to survey temperature irregularities over large areas, thus facilitating indirect studies of ocean currents. We will discuss these methods further in this section.

No matter how detailed and fast the temperature–salinity measurements might be, they cannot completely substitute for direct measurements of ocean flows. Before we discuss basic ideas in current measurements, two principal approaches to the description of fluid motions should be introduced.

First, imagine that an explorer can label a water particle in the sea and observe it for a period of time. By plotting the points through which the particle passes, he will obtain a curve called a **trajectory**. By tracing a great number of trajectories, he will be able to draw a complete picture of the distribution of currents in the region.

An alternate approach would be to place a

23

number of velocity gauges at fixed points, and thus obtain the distribution of current speeds and directions at each moment of time. Instead of a trajectory for each particle, other equally informative curves will be obtained – the **streamlines** – lines tangential to each point of the current velocity vectors at each moment of time. For motion that does not vary with time, the so-called steady-state or stationary current, the streamlines coincide with the trajectories.

The first method is usually called Lagrangian, and the second Eulerian, although, in fairness, it should be mentioned that both were proposed by Euler, although neither of the two great mathematicians ever actually measured currents at sea. The methods now designated by their names were devised on paper for theoretic description of the movements of any fluid. The Lagrangian method provides a picture of the movement of labeled particles, while the Eulerian method gives an instantaneous picture of velocity vectors.

Historically, the technique for recording the co-ordinates of floating bodies in relation to a stationary object, i.e. the Lagrangian method, was developed earlier. A ship seemed most appropriate for doing this. As soon as seamen learned how to find their positions at sea by astronomic observations, they became able to determine the ship's drift off course. From the mid-19th century, ships' logs became a precious source of information on currents. Such data could not be exact, however, since a ship is driven by wind as well as by currents. Besides, the accuracy of astronomic observations is never better than a few kilometers. Nevertheless, most indications of currents on geographic maps are based on information supplied by navigators. Incidentally, it is from navigators that oceanographers adopted the custom of designating currents by the direction *in which* they flow, in contrast to meteorologists who describe a wind by the direction *from which* it blows.

A ship is a very imperfect 'buoy', since it is driven by current and wind simultaneously – not so a sealed bottle. Borne by water alone, it can also carry information itself – an enclosed note stating from where it was launched. Bottle messages, an ancient method for monitoring currents, are still in use. They were probably invented by some ancient mariner eager to pass the word that his ship had been wrecked. This channel of communication used to be considered so important that at one time in England there was a special Admiralty officer responsible for unsealing bottles containing messages. (In *l'Homme qui rit* by Victor Hugo, it is one such officer, Barkilfedro, who opens a bottle that introduces a dramatic change in the hero's life.)

Around the turn of the century, thousands of bottles with notes indicating their initial locations were thrown into the sea. Pulled out in various parts of the ocean, or found on the shore, they provided information on the directions of surface currents (Fig. 2.4).

Research ships, especially small ones, sometimes measure surface currents by the use of a buoy attached to a string. It is an ancient but reliable technique. The ship is moored and a float with a very small above-surface part is lowered into the water on a line. The direction in which the float is driven is the direction of the current, and its velocity is given by the length of the line paid out during a given period of time.

With the development of precise orientation methods and submarine acoustics, the buoy method experienced a renaissance. In the late 1950s, Soviet oceanographers began to measure deep currents with the aid of a standard aviation parachute. The parachute is tied to a rope linking a buoy with a heavy streamlined weight. The parachute is opened at a specified depth, after which the entire set-up begins to move down current, since the drag of the parachute exceeds that of the buoy plus weight. The velocity and direction of the drift are determined by using radio waves and signal lights mounted on the buoy.

Aerial photography of currents also belongs to this class of methods. Parcels containing dye, or colored floats, are strewn from a plane flying along a route and then photographed twice after certain time intervals. This gives rapid measurement of currents over a large area. Currents in coastal waters of the Black and Baltic Seas were investigated by aerial photography in the 1960s. A shortcoming of the method is that it requires a stationary landmark within the viewing field of the lens, which makes the technique inconvenient in the open sea. However, if the plane is replaced by a space satellite, and the dye paths by buoys emitting radio waves, the method becomes quite efficient (see below).

In Western countries, electronic, acoustic and telemetric techniques are applied extensively with the Lagrangian method. Sophisticated neutrally buoyant devices invented by G. Swallow ('Swallow Floats') in 1955 are designed to measure

Figure 2.4 Communication by messages in bottles is ancient and similar methods are still used to measure currents.

deep-sea currents. The device is adjusted to float at a selected density level. Its motions can be traced by detecting sound pulses which are produced by an acoustic transmitter installed in the float. Accurate navigation techniques are crucial in establishing the ship's positions for triangulating on the floats. After some modifications, the floats could be heard with an echo sounder within a range of 20–25 km.

A more remarkable technical achievement appears to be long-range floats deployed in the SOFAR (sound fixing and ranging) channel, an acoustic waveguide layer in the ocean, produced by a combination of pressure and temperature effects on the speed of sound. The possible range of tracking is about 2000 km, with relatively low-energy transmitters mounted on a float. Land-based listening stations are used in this method (Fig. 2.5). Twenty such SOFAR floats were released at a depth of 1500 m during the MODE-I (the Mid-Ocean Dynamics Experiment) research

study in 1973. Many of them were still being successfully tracked late in 1975.

Lagrangian methods proved particularly efficient after it became possible to take read-outs from floating objects from onshore stations and satellites (Fig. 2.6). A large series of automatic buoys were tested, and provided valuable data on the motion of surface ocean layers.

However, far more methods and instruments have been designed using the Eulerian principle. First of all, there are rotary mechanical devices based on the concept of the Ekman–Merz current meter (Fig. 2.7). The velocity is measured by counting the revolutions of a varied impeller and the direction is recorded by a compass. This instrument was still in use in the USSR at many coastal oceanographic stations in the 1970s. The instrument is lowered into the water and lifted up for each reading. To avoid this inconvenience, various techniques have been tested. In particular, cumbersome devices have been built which have

25

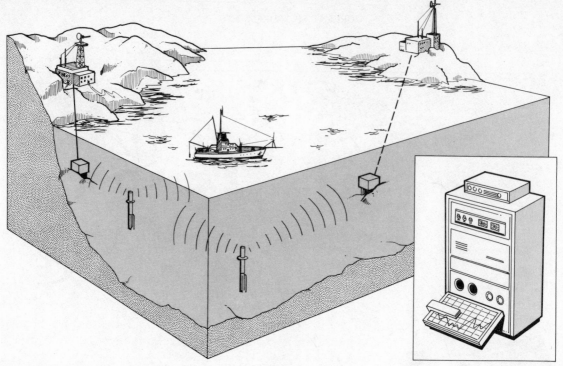

Figure 2.5 Tracking SOFAR floats from shore-based listening stations.

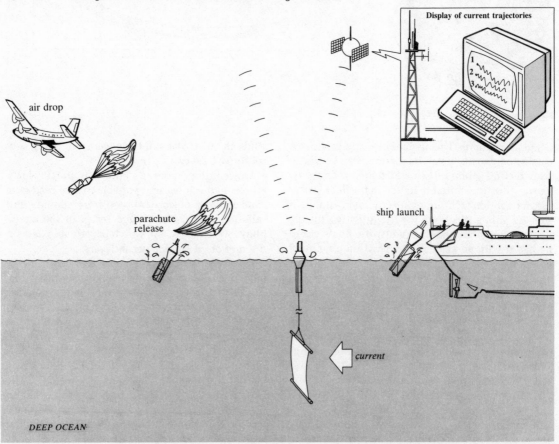

Figure 2.6 One of the space-based systems, COSRAM (Continental Shelf Random Access Management System) to measure ocean currents via satellite.

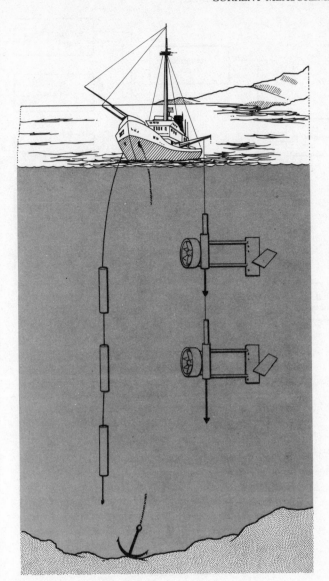

Figure 2.7 Current measurements using an Ekman current meter from a moored ship.

an automatic revolution recorder inside a water-proof case. Up till now, letter-printing current meters have been extensively used in the USSR. Placed at different depths with an autonomous buoy station, they register, on a narrow paper tape, the current velocity and direction at intervals of 5 to 60 min over periods of time from 5 to 30 days. Tedious manual work is required to process the information and to make use of it on a computer.

In the West, such purely mechanical devices have all but been discarded. The latest break-throughs in electronics and telemetry are incorpo-rated in Eulerian-type designs to meet enhanced accuracy and efficiency requirements of current measurement. The wide use of tape recorders and batteries has allowed high-density information storage and lightweight current meters. A wide spectrum of instruments has evolved in the past 20 years.

Although many of these current meters have various design principles, most have similar per-formance characteristics. Following Woodward (1972), a modern current meter constitutes a sys-tem which can be separated into three subsystems:

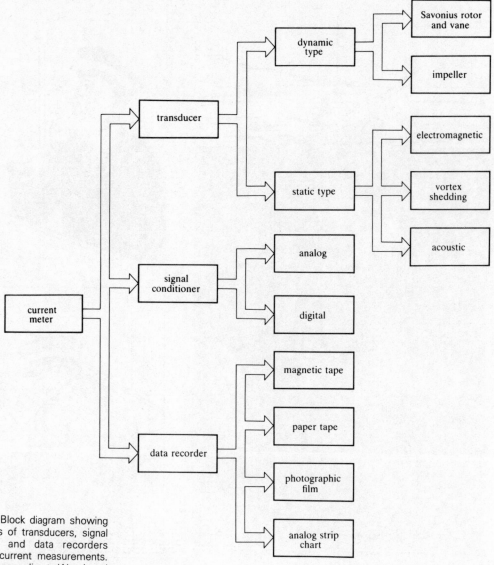

Figure 2.8 Block diagram showing various types of transducers, signal conditioners and data recorders for modern current measurements. (Constructed according to Woodward 1972.)

transducing, signal conditioning and data recording (Fig. 2.8). It is worthwhile to describe modern transducers that convert the water velocity vector into some form of energy, usually electrical energy. There are two major categories of transducers in use, dynamic and static. The former is designed to be set into motion that is proportional to the water velocity vector. The static transducer remains stationary and measures some property of the water motion related to its velocity vector.

Among the most popular dynamic transducers is the Savonius rotor and vane, invented in 1925 but not used extensively before 1950 (Fig. 2.9).

The device is not sensitive to vertical motion since its rotor of two hollow half-cylinders is mounted on a vertical axis. It produces a large torque even in small currents. The current direction is measured separately with a vane electrically coupled to a magnetic compass.

An impeller-type sensor (Fig. 2.9) rotates about an axis that is in line with the water flow. Thus the entire system or the sensor itself is freely suspended to provide proper alignment with the flow, and a magnetic compass is referenced to the body of the suspended device in order to measure the current direction. In both instruments, rotation is

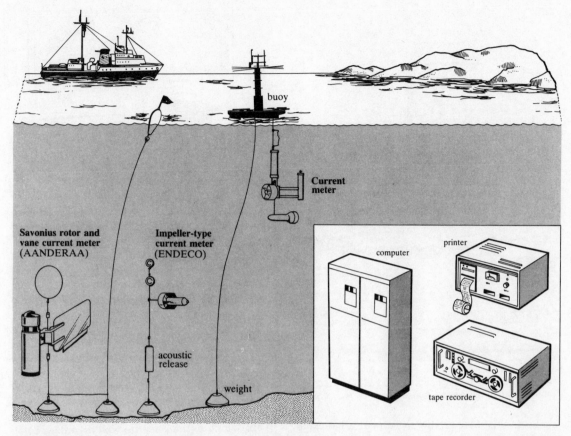

Figure 2.9 Current meter mooring arrays: (1) subsurface deployment with a Savonius rotor-type current meter; (2) subsurface deployment with impeller-type current meter and acoustic release; (3) buoy station with freely suspended current meter. Also shown are the printer, recorder and computer.

sensed by the switch and by magnets, although the signal may be transmitted to a gear located inside the instrument housing.

A static transducer has an advantage in that it avoids mechanical motion in the device and thus eliminates inertia problems. The most common static transducer is an electromagnetic device (the Geomagnetic Electrokinetograph, GEK) based on the principle that a voltage is induced in a conductor set in motion through a magnetic field (Fig. 2.10). The ocean current is a moving conductor, and the magnetic field is either the Earth's field or an applied field. Another method, the so-called vortex-shedding technique, is based on the modulation of an ultrasonic beam passing across a number of vortices behind a round cylinder that breaks the flow. This acoustic method uses the difference between the speed of sound transmitted downstream then upstream.

The signals produced by a transducer are accepted by the signal-conditioning subsystem which, in turn, outputs them in a form suitable for recording. It is beyond the scope of this book to describe electronic signal conditioners and recorders. However, it should be mentioned that these systems are usually designed to sample current vectors at certain discrete intervals, thus permitting long-term observation even with limited data storage capacity.

Current studies made with instruments lowered from a moored ship or hung from a moored buoy involve substantial amounts of time and labor. Moored current meters can give only a time series at a few selected depths, and some narrow jets may not be recorded. Designers are therefore looking for ways of building passively drifting current meters. One such unit, the passive free-fall instrument, measures the velocity structure of an ocean current. The principle of this device is similar to the electromagnetic method mentioned

Figure 2.10 Electromagnetic method (GEK) in current measurements. The current component normal to the line connecting two electrodes is measured and recorded on board (*inset*).

above. In this method, the instrument is ballasted to fall to a given depth, or to the bottom, and then return to the surface. As a result, a detailed vertical profile can be obtained. Several principles in the velocity profiles that have been recently employed are described in detail by Baker (1981).

By the 1970s, observational oceanographers had an impressive variety of moored current meters and float techniques to measure ocean currents with high accuracy and over long periods. Even the first tests of new instrumentation in the open ocean planted seeds of doubt as to the well balanced picture of ocean circulation described in Sections 2.2 and 2.3. For instance, it was found that ocean movements outside boundary flows were not as sluggish as the general picture predicted. Sophisticated observational techniques demonstrated that the instantaneous state of oceanic motions might be radically different from the time-average current field. The first long-term moorings and float trackings displayed new circulatory phenomena, which, for the sake of convenience, were named 'eddies'. These are large-scale (more than 100 km) fluctuating currents, which in time acquire well organized rotating patterns ('rings').

The discovery of eddy-like motions in the open oceans substantially affected the strategy of oceanographic observations. The design of field experiments became a key problem in studies of large-scale oceanic processes. Given the high cost of both equipment and logistics, only a modest number of moorings and very few research vessels can be operated simultaneously. It has become obvious that constant reconnaissance information on large-scale dynamics is a major key to success in any field experiment with limited observational means. Without crude pictures of large-scale motions, the transient phenomena, such as eddies, can be missed or inadequately sampled during the observational period.

The difficulties faced by oceanographers in obtaining any fast surveys of oceanic movements are easily understood. Unlike meteorologists, oceanographers do not have a global synoptic weather network, accumulating data routinely. The oceans, by their very nature, present considerable difficulties to fast observations: in the opaque ocean environment no balloon (or float) can be monitored by visual or radio tracking. Oceanographic research vessels are too slow and labor-intensive to implement fast reconnaissance studies over large areas, and unmanned platforms and buoys are so expensive that they cannot be maintained for a long time merely to supply background information.

The advent of space technology has opened up the possibility of investigating the ocean on a very large, indeed global, scale. This became possible because of the rapid development of remote sensing of the ocean surface, which brought novel information on surface temperature and color, surface topography and wind stress. This information was gradually assimilated into oceanographic research and has already become an indispensable component of any large-scale ocean study.

Two satellite techniques are directly related to ocean circulation. The first is high-resolution thermal infra-red imagery, in which an infra-red radiometer measures the intensity of radiation emitted from the sea in the infra-red band in a broad swath beneath the spacecraft. Upon calibration with conventional observations, these measurements yield estimates of sea surface temperature. Many pictures like the one shown in Figure 2.11 have been widely published. Some such infra-red images have allowed researchers to track movements of frontal zones in the ocean, to identify warm and cold eddies in the boundary currents, etc.

Taken in isolation, the satellite-based temperature maps have relatively little value for the following reasons. First, the temperature of the 1 mm layer of the ocean sensed by a radiometer is not necessarily related to dynamic phenomena – it may reflect various processes at the actual sea surface (transfer of heat, moisture, formation of foam, etc.). Kuftarkov *et al.* (1978) demonstrated that this skin layer is remarkably stable, even during a 10 m s^{-1} wind, and preserves the vertical temperature difference of about 0.4–2°C for a long time. Secondly, large portions of the ocean are frequently covered by clouds, which block infra-red radiation from the sea. Without the support of direct measurements, the interpretation of some infra-red images is a difficult task.

Another important development will make remote sensing a more important instrument in large-scale monitoring of subsurface temperatures: it is the laser Raman backscattering technique that enables one to record temperatures to a 10 m depth with an accuracy of ±2°C. It is expected that the system will work to at least a depth of 60 m with an accuracy of ±1°C.

Another satellite technique is the measurement of the surface topography by satellite altimetry. An altimeter, a pencil-beam microwave radar, measures distances between the spacecraft and the Earth with an accuracy of a few centimeters. When the ocean is at rest it assumes a shape called the **geoid**; this resembles an ellipsoid of rotation that bulges out somewhat at the Equator, but has non-uniformities of about 100–200 m in height due to the variable gravitational attraction of various parts of the Earth's crust. Ocean currents cause the sea surface to depart from the geoid. These relatively small elevations and depressions (no more than 1–2 m) should be detected by the altimeter while the satellite passes over many ground tracks such as those shown in Figure 2.12a. The altimeter fixes the sums of geoid and current-induced non-uniformities relative to some reference ellipsoid. An example shown in Figure 2.12b reveals mostly geoidal forms. In order to isolate the dynamic components of surface topography, the geoid should be subtracted from the total height. Then the topography of the ocean surface is indicative of some known features of the ocean circulation. On the cover such a topography map highlights the Gulf Stream system by sharp surface slopes along the core of the current. (The relationship between the geostrophic current and surface topography is discussed in detail in Section 3.4.)

Currently, there are several other satellite-based observational techniques in operation, which are not discussed here. The reader is referred to Apel (1976), Szekielda (1976), a special issue of *Oceanus* (1981) and Baker (1981) for more comprehensive reviews on space oceanography.

Large-scale monitoring has also become possible due to an acoustic technique of great potential proposed by Munk and Wunch (1979). It is named 'ocean acoustic tomography' after the medical procedure of producing a two-dimensional display of interior structures from exterior X-rays. The technique monitors the ocean's density structure that is inferred from the acoustic time travel between a number of moorings. The method will eventually describe the velocity distribution over large water areas with sufficient accuracy. The system will be very cost-effective with respect to the results of large-scale monitoring that it will be capable of producing.

The introduction of space techniques into

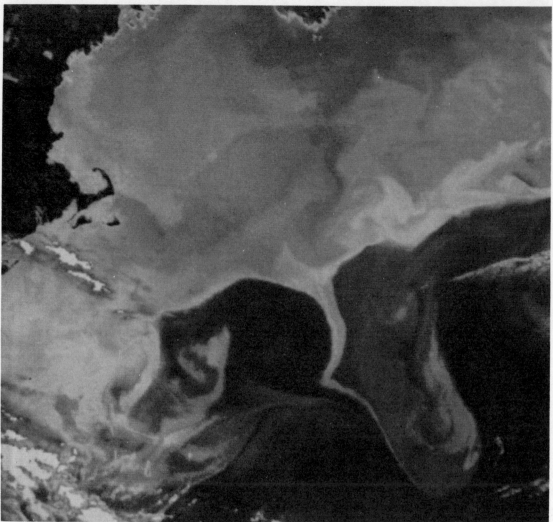

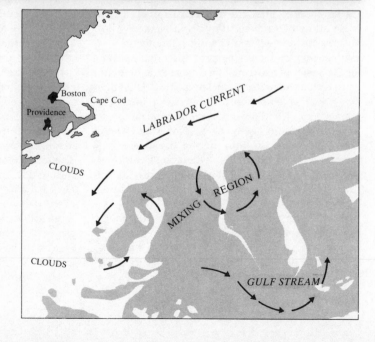

Figure 2.11 Infra-red satellite image of the warm Gulf Stream's northern edge and the cold Labrador Current at 1400 GMT on May 23, 1978. Warm water (dark) is mixed up with cold water further north (white). Puffy patches at the lower left and south of Cape Cod are clouds. Reproduced by courtesy of R. Legeckis, National Earth Satellite Service.) Locations of currents and mixing region are shown on a separate sketch.

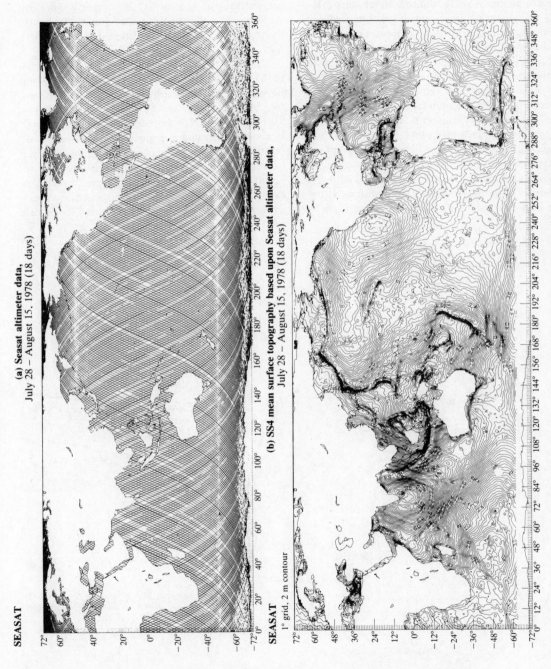

SEASAT

(a) Seasat altimeter data,
July 28 – August 15. 1978 (18 days)

(b) SS4 mean surface topography based upon Seasat altimeter data,
July 28 – August 15, 1978 (18 days)

SEASAT

1° grid, 2 m contour

Figure 2.12 (a) Ground tracks of passes of GEOS 3 and (b) mean sea surface topography computed from SEASAT altimeter data. Contour interval is 2 m. The surface is dominated by geoidal undulations, but topography caused by ocean circulation is included in the altimeter measurements.

oceanographic research allowed the better design of ocean experiments and observational strategies in order to investigate intricate mechanisms of oceanic phenomena. A marked historical evolution has taken place in oceanographic research with the accumulation of data on ocean behavior. In the early days, the expeditions were multidisciplinary and multiregional, yielding basic information in the form of maps and atlases. Nowadays, oceanographers focus their attention on specific phenomena, such as boundary currents, the mid-ocean environment, deep-ocean flows, etc., using specially designed experiments. But only recently, in the 1970s, have oceanographers addressed decisively the problem of nearly continuous monitoring of such transient phenomena as eddies, internal waves and the temporal structure of the ocean. This has become possible as

satellites, buoys, moorings, drifter technology and ships have been integrated into a composite system, the satellites being used to read out information from the buoy network, and to monitor the sea surface and document the changes in the ocean environment, whereas traditional conventional means were unable to do this.

The first two experiments – the 1970 Soviet POLYGON-70 and the 1973 US–UK MODE-I – can be regarded as pilot studies based on conventional techniques: ships, moorings and temperature–salinity profiles. The experimental sites and duration of measurements appeared to be insufficient for the adequate description of eddies passing through the mooring arrays. Measurements of ocean variabilities raise far more questions than they have been able to answer. However, these experiments constituted a background for future

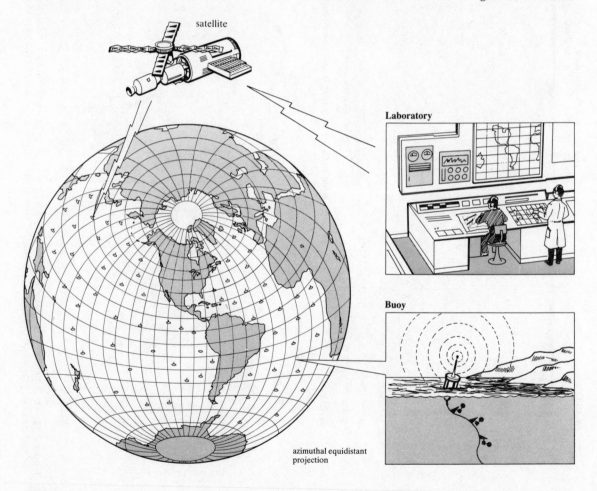

Figure 2.13 In the future all the components of 'oceanic weather' will be recorded by an international network of buoys interrogated from satellites and ground stations with subsequent data processing at computer facilities.

endeavors, providing information on spatial and temporal scales of oceanic phenomena and tests of available measurement techniques.

The experiments that followed (the international POLYMODE experiment, studies of the Gulf Stream rings and others) have been based on better experimental designs. The organizers of these studies adopted a very flexible strategy of tracking major circulatory patterns using a combination of long-term mooring arrays and synoptic surveys based on the XBT profiles and SOFAR float tracking. In later stages, the position of the Gulf Stream has been monitored from satellites, and individual rings have been detected at the points of their formation, to be then followed during their propagation. The reader is referred to Sections 5.7 and 6.6 for some results of these experiments.

It is these local studies that have been the primary occupation of the oceanographic community in the past 10–13 years. However, much effort was involved in instituting a global monitoring system to document data on the state of the ocean.

The backbone of such a system is satellite-based read-out from a network of autonomous buoys, which operates similarly to a network of weather stations. The initial US project was to build a telemetric system of monitoring and reading out data from 540 buoys, including 261 to be placed in the deep-water parts of the oceans. Because of the enormous cost and importance of this project, other nations have to participate. In particular, Britain, the United States, Canada and West Germany have participated in a unique scheme of information exchange. The system is supposed to be a first step toward the creation of an International Global Ocean Stations System (IGOSS), to be launched jointly by the International Oceanographic Commission and the World Meteorological Organization. IGOSS is to become the oceanographic equivalent of the World Weather Watch (Fig. 2.13).

Further development of this system is visualized as co-operation between the West and the East and the creation of international co-ordination and computing centers for global collection, processing and transmission of data on dynamics of the ocean and the adjacent atmospheric layer. The major elements of this monitoring system are: (1) satellite observations of the surface temperature and surface pressure; (2) direct measurements of the surface currents by drifters and a modest number of moorings; and (3) deep-sea monitoring by combinations of acoustic tomography, ship-borne hydrography and moorings. All elements are now present or are being tested; we could have such a system in the next ten years (Baker 1981).

2.6 Mapping of ocean currents

Progress in mooring and drifter technology provides oceanographers with ever-increasing amounts of current measurements over large areas and at different depths. Since data are accumulated continually, it is extremely important to summarize this flow of digital information and digest it for use by science and industry. An urgent need has arisen for standard procedures that would reduce the data on currents to a form convenient for survey and analysis. As far as mapping of currents for practical purposes is concerned, such procedures were developed long ago. However, before describing them, let us once again take a brief look at the possible requirements of the various users of this information.

Navigators are usually interested in mean data on transport in the surface layers of the ocean. For them it is sufficient to get an indication of the currents averaged over a longer period of time, such as one day. This is why current maps based on last-century data on ship drift or bottle messages are still quite satisfactory. However, near ports, channels and straits, where currents are affected by both large-scale oceanic motions and tides, navigators need more detailed information. Guides and manuals therefore indicate the variations of currents during the tide.

Fishermen need data on places of divergence and convergence, the regions of upwelling, and also on current velocities in such areas.

Port builders and ocean engineers are concerned exclusively with the nearshore zone where currents may clog channels with sediments, undermine or otherwise affect rigid and floating structures. They need data on maximum current velocity and duration. The same is true for cable laying and other underwater work. These practical workers thus mainly need data on average maximum currents and their occurrences and durations in time.

In recent years, those concerned with the problems of sewage and industrial waste discharge into the sea have also begun to require detailed information on the structure of currents and their small-scale variations. In such cases, it is essential to know the regions and the time of action of weak currents that slowly scatter impurities which may pose a threat to the sea inhabitants. They are also interested in currents in order to conduct safe disposal of wastes into the coastal zones where strong currents can further dilute concentrated plumes. It is possible to predict the fate of industrial effluents and concentration of pollutants in the sea after studying the small-scale irregular motions, called turbulence (see Section 3.6). With the growing importance of environmental protection, it is becoming ever more essential to collect data on currents in potential discharge areas.

We shall begin with a look at how averaged currents are shown on maps or charts. First of all, the data are averaged over the surface area. Average velocities and current directions are calculated from all the measurements taken in each small square of the ocean. If the number of observations in the square is large, it is also possible to determine the probability of occurrence of the current of a given strength in each direction. This is illustrated by Table 2.1.

On the left-hand side of the table, equal intervals of current velocities are selected. Along the top, groups of current directions are written as degrees of deflection from one of eight principal directions. All the observations in the area are tallied according to current velocity and direction. The number of observations falling into any separate box shows the frequency of any current vector in this range of velocities and direction. Then the frequencies (in numbers of observations or as a percentage of the entire set of data) are plotted on a chosen scale on a polar diagram (Fig. 2.14a) and isopleted. This graph distinctly shows the most frequent directions and velocities in an area.

Usually, a less precise method is employed. The weighted average velocity is found for each direction, and the average percentage of occurrences for this direction is calculated (see bottom two rows in Table 2.1). Then the average vectors are drawn on a polar diagram with the occurrence in percent shown by the length of an arrow (Fig. 2.14b). Several such frequency diagrams displayed on a map give a representation of the typical distribution of currents in the region under study (Fig. 2.14c).

Another method of generalizing instrumental observations of currents involves plotting a diagram of the actual movement of a particle. The current vectors are plotted in a time sequence – the end of one being taken for the beginning of the next vector (Fig. 2.15a & b). As a result, the average vector is calculated, and the current variation is estimated by using a quantity called **stability**.

Stability B of an ocean current is defined as the percentage ratio of the length of the average vector $|\mathbf{v}|$ to the arithmetic mean velocity v. Thus

$$B \ (\%) = 100 \ |\mathbf{v}|/v$$

Table 2.1 Two-way frequency distribution of current velocity and direction.

Velocity \ Direction	N (338–22°)	NE (23–67°)	E (68–122°)	SE (123–157°)	S (158–192°)	SW (193–247°)	W (248–292°)	NW (293–337°)	Total
0.00–0.05	2	3	0	3	0	1	0	1	10
0.06–0.10	11	17	5	16	15	17	2	8	91
0.11–0.15	32	36	21	16	11	27	12	23	178
0.16–0.20	44	49	31	22	6	32	11	29	224
0.21–0.25	69	82	40	9	2	11	2	31	246
0.26–0.30	21	83	25	9	0	2	3	12	155
0.31–0.35	20	62	16	1	1	1	0	9	110
0.36–0.40	9	28	10	0	0	0	0	3	50
0.41–0.45	5	11	1	2	0	0	0	2	21
0.46–0.50	2	1	1	0	0	0	0	1	5
total	215	372	150	78	35	91	30	119	1090
average velocity (m s⁻¹)	0.24	0.33	0.23	0.77	0.12	0.16	0.16	0.20	
occurrence (%)	20.1	34.1	13.7	7.1	3.2	8.3	2.7	10.9	100

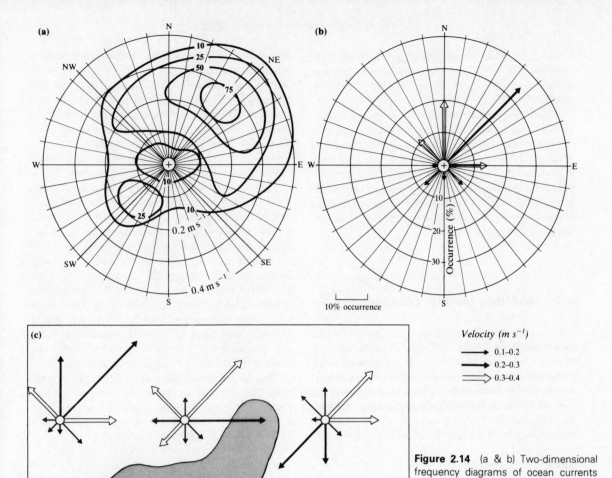

Figure 2.14 (a & b) Two-dimensional frequency diagrams of ocean currents (for explanation, see text and Table 2.1) and (c) map showing several such diagrams (in conventional scale).

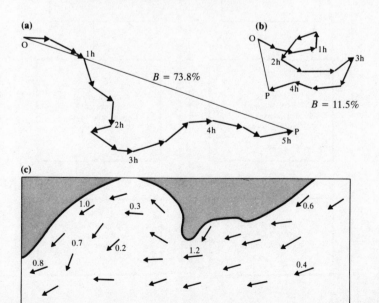

Figure 2.15 Successive-vector diagrams for (a) steady and (b) unsteady currents, and (c) a map of average vector field.

Figure 2.15a illustrates an instance of a rather high stability ($B = 73.8\%$), and Figure 2.15b an instance of random motion with a very low stability ($B = 11.5\%$).

When it is possible to observe currents at many points at sea for a long enough period of time, and determine the average vectors, an easily readable map of currents can be compiled, such as that shown in Figure 2.15c. By varying the direction, thickness and length of an arrow one can represent graphically the current stability and velocity. Instead of arrows, probable trajectories of water motion can also be shown.

2.7 Modeling oceanic phenomena

New exploratory observations in the form of local experiments (MODE, POLYMODE and others) clearly demonstrated that the instantaneous dynamic state of the ocean, at least in its upper strata, largely departs from the classic picture of ocean circulation depicted above (Sections 2.2 and 3). The transient phenomena (eddies, vortices, meanders, etc.) are ubiquitous, and have compar-

able energy to the principal ocean currents. Does this mean that the historical data bank on currents does not represent the real situation in the ocean? Not at all. It is simply that the old methods provided the dynamic picture averaged over large intervals of time, whereas new instruments, with very high-frequency response, display unknown time-dependent motions representing 'instantaneous' pictures of the ocean circulation.

These two presentations of the same current field are not mutually exclusive. Each of them reflects certain realities, but has a different causal structure and energy source. In order to understand the basic mechanisms shaping both the organized average circulation and the highly irregular eddy structures, a certain universal analytical tool can be applied. It is a **conceptual model**. In reality, modeling is nothing other than a number of creative and purely intellectual acts to examine and present, in easily observable form, the processes or phenomena under study and make predictions of their future behavior.

Though models differ widely depending on the scientific discipline, research methods, investigator's skill and background, a certain set of procedures can be usually traced in a model's design and subsequent analysis (Fig. 2.16).

Model:

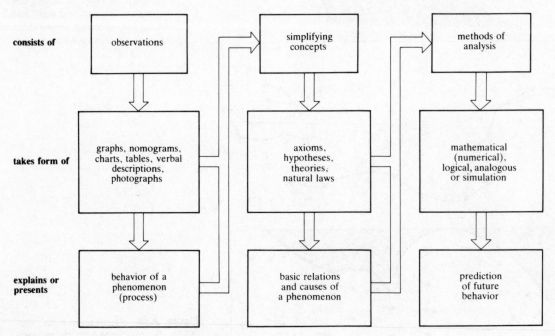

Figure 2.16 Block diagram of modeling procedures.

Any modeling starts from observations which adequately describe the phenomena under study. In dynamic oceanography these data are presented in the form of temperature–salinity fields, T–S diagrams, current maps (trajectories or streamlines), topographies of the sea surface or other surfaces in a water column, tables of characteristics, etc.

But no matter how representative and precise these data are, they do not in themselves explain the causes of phenomena. Some essential simplifying concepts are needed – often called axioms, hypotheses, theories or natural laws. The essence of each model is skilful application of these concepts to describe basic mechanisms controlling the behavior of a phenomenon. Many of these simplifying concepts will be familiar to the reader. They are axioms of Euclidean geometry, balance of forces and amount of substance in a given volume, physical and chemical processes, energy transformations, etc. The quality of a model depends on a modeler's ability to isolate the basic controlling mechanisms and flows of substances in the environment under study to reveal the causal structure of a phenomenon.

But formulation of basic relations alone will not complete the model's design. A number of procedures are still required to analyze the processes under study and make predictions. Depending on the nature of a phenomenon, these procedures are logical, mathematical or involve some other easily observable behavior in order to simulate the process under study (this last model is sometimes called analogous or **simulation**). In environmental disciplines (oceanography, meteorology, hydrology, ecology, etc.), mathematical models are most frequently used. In order to stress that numerical information is to be obtained with the aid of computers, the terms **numerical model** is applied. However, until recently the basic mathematical models were relatively simple and could be solved without computers. These are called **analytical models**. For educational purposes, the majority of textbooks deal primarily with analytical models because they present oceanographic processes in the most simplified and easily observable form.

In Section 1.4 the reader had an opportunity to see how a logical model was applied to describe the atmospheric wind patterns on a non-rotating and then rotating uniform globe. A familiar physical process (convection) and simple balance of forces (horizontal pressure gradient vs. Coriolis force) were analyzed to explain the three-cellular

circulation in the atmosphere (Fig. 1.4b). This method will be largely used throughout this book to describe the average large-scale ocean circulation. However, the reader should realize that behind the logical structure of each model is hidden a substantial mathematical apparatus. It surfaces only in the simplest cases.

The nature of the ocean environment imposes some limitations on models of ocean currents. The oceans are spread over the planet's surface in a peculiar manner and have a complex bottom topography. All oceans are connected and there is constant water exchange between them. The external driving forces – wind and thermal gradients – are also distributed extremely unevenly. And, as if to complicate the picture further, the entire system rotates with the Earth. The structure of oceanographic fields is also complex, as shown, for instance, by zones of convergence and divergence due to vertical motions. From the previous description we know that oceanic movements are multiscale. Apart from 'lawful' behavior in the form of the average large-scale circulation, there exist various swift and essentially random circulatory patterns. These latter phenomena do not have an apparant cause, and appear as a result of the combination of many factors (hence, the surprising discoveries of eddy-like motions during MODE and POLYMODE experiments).

How can mathematical models portray and analyze these complex motions? One would assume that some filtering technique is required to isolate the slowly changing average large-scale circulation from the rapid small-scale movements of random nature. Such a technique does exist and can be applied to the observed circulatory fields. However, in modeling procedures, the scaling problem is not the major issue. A modeler of ocean currents usually specifies the scale of a phenomenon from the very outset, whereas small-scale processes are counted as energy sinks (or sources) via several empirical co-efficients. The opposite problem, where most attention is focused on small-scale processes, with large-scale flows known from observations or otherwise, is now an active field of research. In this book we primarily deal with the first type of modeling: the large-scale fields of current, density (or temperature and salinity) are affected by external factors – the wind and heat fluxes through the ocean surface.

Since very limited mathematical symbolism is used here, the basic relations are described verbally. Mathematical models of ocean currents are usually based on the equations of fluid mechanics. These equations express the universal physical laws as applied to fluids: Newton's second law of motion, the law of conservation of mass, the conservation of energy, the conservation of angular momentum, and the laws of thermodynamics. Each of these laws is extensively used in dynamic oceanography. The second law of mechanics – the balance of forces – is stated in the form of **momentum equations** linking the three components of current velocity (along the axes of a Cartesian coordinate system) with density and pressure. The conservation of mass results in another relationship, the **equation of continuity**, which links the velocity and density fields. The laws of thermodynamics allow us to connect density with temperature, salinity and pressure in the form of **equations of state**. A model is called 'closed' when the number of equations is equal to the number of unknown functions.

When there are not enough dependences, or when some new, unknown values have to be taken into account, one must reconsider the physical picture of oceanic phenomena to derive new equations, or to find the relations between the quantities empirically, from observations.

In general, the models describe time-dependent oceanographic fields which develop under the action of external forces. However, it has been proved to be sufficient to explain the nature of ocean currents in a non-accelerated frame of motion, which does not change with time. Such processes are called **stationary processes**.

Knowing the equations that describe a process is not sufficient, however, to construct a model of it. The behavior of fluid in a basin is affected by the conditions at the basin's boundaries, because the current and density fields in the ocean are influenced by external factors such as wind, transport of heat and water across the ocean surface, and features of its bottom and shores.

It is necessary, therefore, to formulate mathematical relationships – the boundary conditions – that describe the regularities at the boundaries of the basin or of the ocean area being studied.

The need for boundary conditions follows from the essence of the equations of fluid mechanics. In fact, these are not simple algebraic relationships, but differential equations. With some approximation, it can be said about such equations that they link not quantities, but their differences, the 'specific' variations per unit of time and space.

This suggests that they can be used to calculate the field of an unknown value if that value is defined at the boundary.

Boundary conditions sometimes follow from obvious facts such as that water will not flow through a solid boundary. Other conditions are not exactly true but are adopted as hypothetical assumptions. Fluid dynamics equations as such require a certain number of boundary conditions to be solved. For this reason, in addition to boundary relations substantiated by physics, one sometimes has to introduce certain hypotheses about the boundaries, such as the standard assumption that the current velocity exactly at the sea floor is equal to zero.

Transformation of all fields depends on the initial state assumed in the study. This means that some initial conditions should be set in advance, i.e. all the fields must be defined at a certain moment of time at all points of the space being studied. For a stationary model the initial conditions are immaterial since, by definition, the process is not dependent on time.

Once the laws of motion are expressed in the form of equations and boundary conditions, the construction of the model is virtually complete. If the model is built correctly, the problem will be correct, that is, there will be enough equations and boundary conditions to determine all the unknowns of the field. All that remains to be done now is to solve the mathematical problem and interpret the results, which also requires much effort and completes the cycle of the modeling study.

Unfortunately, the differential equations used in fluid dynamics models still lack easy solutions. This means that in practice it is not feasible to derive an exact expression, for instance of an algebraic relation, to link the knowns to the unknowns. What one can do is to try to make use of the inherent flexibility of a mathematical model – reformulating the problem in such a way that the contributions of various factors to currents and other fields would not be considered at all once, but in a sequence of increasingly more complex phenomena.

With the advent of modern numerical methods and high-speed computers, increasingly complicated models of ocean circulation are being formulated and solved. Modern models take into account factors such as the complex wind field over the oceans, the fluxes of heat and moisture across the ocean surface, and the shore and bottom geometry of the real basins. Spectacular results have been obtained, reproducing intricate details of the global ocean circulation using numerical models. However, the basic mechanisms of ocean currents can be explained with the aid of analytical models.

In the rest of this book, increasingly complicated mathematical models in the form of qualitative descriptions will be used to explain and present graphically physical mechanisms of ocean currents. Small-scale phenomena will be also taken into consideration, when possible, in the form of cumulative effects upon the average current structures. However, large energy-containing patterns, such as eddies and the Gulf Stream rings, do not fit our frame of analysis. The causes of these phenomena are not yet known, and therefore they will be treated differently.

3 Causes of ocean currents

3.1 Evolution of the theories of currents

In the foregoing analysis of the ocean and the atmosphere as two interacting subsystems, we have identified two major energy inputs into the ocean. These are the wind stress over the sea surface and heat fluxes in the form of latent heat, conductivity and solar radiation. Is it possible to tell in advance which of the two factors – heat or wind – is mostly responsible for inducing the movements that occur in the ocean? Analogously to the atmosphere, one might be tempted to assume that masses of sea water move under the effect of temperature gradients. But then the atmosphere is heated from below, whereas the sea is heated from above – and we know from our everyday experience that the best way of heating a volume of air (or water) is from below.

The depth to which the ocean is heated is extremely shallow compared to its total depth. Daily fluctuations of sea temperature are hardly noticeable at 15–20 m, and below 150–200 m even seasonal fluctuations are hardly discernible. The thermal 'breathing' of the sea is thus very shallow. However, in the polar regions the ocean is cooled at the top, but is the ensuing convection just as efficient as heating from below? Apparently not, because the eternally cold waters of the ocean abyss have become chilled at some time in the past near the Poles (we shall return to this later). The fact that in the course of many years the temperature below 1000 m does not undergo noticeable changes, maintaining the remarkable uniformity of ocean waters, shows clearly that heat fluxes cannot be the main source of energy setting the surface ocean waters into motion.

In contrast with the atmosphere, which is only affected by heat, the ocean is subjected to the action of a second driving agent – the wind. How deep does wind energy penetrate? Jules Verne's Captain Nemo already knew that a plunge to a depth of 20–30 m shelters a submarine from the most ferocious storm: sea waves are hardly ever felt at that depth. Can currents caused by storm winds exist at such depths? Generally, can wind systems produce strong enough mechanical signals in the water body, such as do major oceanic gyres? Again, judging by everyday experience, the wind does not seem to be a particularly stable force.

The issue of what factor is principally responsible for ocean currents has been debated among oceanographers for over 50 years. Few authors have ever taken a definitive stand in support of either hypothesis. It has been a cautious and baffling struggle, with opponents sometimes voicing counterarguments a dozen years after the original contention. They would often formulate an argument more lucidly than its originators, only to destroy it point by point on the following page. Then, fresh data would be brought in by a research ship, and the old theory would be revived. Rather than a conflict of two opposing schools of science, this controversy was a slow process of evolution of various views on the problem as more and more specialists – geographers, physicists and mathematicians – joined in its investigation.

Geographers were the first to formulate their views on the nature of currents and these views were to remain predominant until the 1940s. Among these were eminent students of the ocean, such as O. Krummel, B. Helland-Hansen, F. Nansen, G. Schott, S. O. Makarov, Yu. M. Shokalsky, G. Wust, A. Defant and others. Numerous expeditions and observations conducted over the years provided them with enough data to describe the basic features and patterns of ocean currents.

At that time, no method existed that would allow determination of the specific contributions of the individual factors to the generation of currents. Hypotheses were mostly intuitive. For instance, geographers presumed that the forces producing the huge subtropical gyres were the slowly varying climatic factors, such as cooling and heating of the sea and unequal salinity of the water due to evaporation, precipitation, and the freezing and melting of ice.

What about the wind? It was believed to perform

an auxiliary function as a temporary factor capable only of slightly modifying the behavior of currents in the surface layer of the ocean. Yet, as the reader will recall, each subtropical gyre includes a section of trade-wind currents, which are obviously associated with the prevailing winds, and there could be no doubt that winds were responsible for currents in that part of the ocean at least.

As time went on, physical methods were introduced into ocean studies and it became possible to evaluate the contribution of each individual factor.

V. Bjerkness and V. W. Ekman were probably among the first scholars to analyze currents by purely mathematical techniques.

The Bergen school of fluid dynamics, founded by Bjerkness, completed the transition of physicomathematical methods in studies of the weather and the sea. Based on Bjerkness's circulation theorem, the first oceanographic method used for calculating currents by density fields was developed and later became known as the **dynamic method**.

In 1905, Ekman created the first model of wind-driven currents. He discovered that wind energy would be sufficient to generate currents with a speed close to that of real currents. At that point, however, it was too early to revise the 'climatic' concept, for Ekman only succeeded in evaluating the wind-driven currents for very simple cases. In 1933, G. Goldsbrough demonstrated that climatic factors could only produce unrealistically weak currents in the World Ocean. However, the prevailing oceanographic thinking of the time was cautious in adopting this revolutionary view, regarding this model as complex and too academic.

It was not until the appearance of works by the Russian, V. B. Shtockman (1946), and the Americans, H. Stommel (1948) and W. Munk (1950), that the 'windward turn' was finally made, and the major subtropical gyres and the system of equatorial currents were simulated theoretically in terms of the wind stress alone, without recourse to climatic factors. These models appeared to be so convincing that a general infatuation with the wind-current theory swept through oceanography, reaching a climax in the mid-1950s. In scores of papers, the climatic factor was all but ignored and the wind was interpreted as the only source of oceanic circulation. This drastic view, however, had again to be abrogated after more knowledge was gained about the ocean's depths.

It was discovered that currents exist at all depths – even in layers sheltered from the direct impact of wind energy. Besides, the wind factor, taken alone, failed to account for the temperature and salinity fields associated with the currents. P. S. Lineikin (1955) first renounced the lopsided 'eolian' approach to oceanic circulation. He developed a new basic theory describing the interaction between the current and density fields and offering a new picture of the activity of climatic factors in ocean dynamics.

In the following sections, we shall develop a closer acquaintance with oceanic movements using basic physical and mathematical tools. We will first examine a steady-state, or stationary, movement, with all velocities assumed to be constant in time, provided all the forces acting on the fluid balance one another and inertia is ignored.

3.2 Horizontal pressure gradient

Leave open the door in a room containing a hot-water radiator, wait till a steady regime sets in, and then observe the movement of air at different levels with the help of a candle flame (Fig. 3.1). This experiment is routinely used in classrooms to illustrate thermal convection. The air will flow out of the room near the ceiling, and into it near the floor, owing to the air pressure difference between the inside and the outside at the corresponding levels.

The next experiment brings us back to the sea. First staged by the Italian scholar, Luigi Marsigli, in 1681, it is known in the history of oceanography as the **Marsigli box** (Fig. 3.2). Marsigli was intrigued by a curious observation made by fishermen in the Strait of Bosphorus, a deep narrow channel between the Black Sea and the Mediterranean. They reported that two currents existed in this channel – an upper current from the Black Sea into the Mediterranean and a lower current flowing in the opposite direction. Marsigli explained the phenomenon by using a box divided into two compartments by a vertical partition. At the top and near the bottom, the partition had holes. Marsigli stopped up the holes and filled one compartment with water from the Black Sea, and the other with Mediterranean water that reached to the lower edge of the upper hole – he then opened the holes. The saltier, heavier Mediterranean water started to flow through the bottom hole and the lighter Black Sea water through the upper

Figure 3.1 Experiment with thermal convection maintained by a hot-water radiator. An excess of rising air near the ceiling and its outflow occur because of thermal convection at the window.

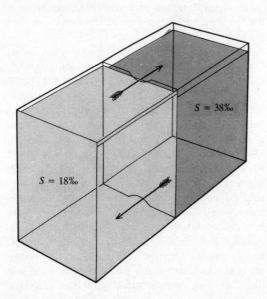

Figure 3.2 Salt water convection in the Marsigli box. Arrows indicate the direction of the pressure gradient and the transport of various water masses until their complete mixing.

hole. After a short time, the two liquids mixed completely.

However, in nature, Black Sea and Mediterranean waters can never mix completely, since the rivers entering the Black Sea dilute it continually and the immense persistent evaporation from the Mediterranean maintains its salt content at a higher level. As a result, the upper and lower currents should be encountered permanently in the Bosphorus. Any traces of doubt that might have remained after this convincing experiment were dispelled in 1885 by direct measurements performed by the eminent Russian navigator and oceanographer, S. O. Makarov. As the captain of the fire ship 'Taman' moored off Istanbul, he conducted regular observations of the sea currents. Filling a barrel with different amounts of sand so that it would float at different depths, Makarov recorded the direction in which the line tied to the barrel was deflected. When submerged, the barrel invariably drifted toward the Black Sea, sometimes with a speed of 1 m s^{-1}. Incidentally, this perfect confirmation of the Marsigli hypothesis earned Makarov an honorary prize from the Russian Academy of Science 'for studies in hydrology of sea channels'.

These phenomena, observed under laboratory conditions and in a real sea channel, bring us closer to understanding a major force acting in the ocean – the **pressure gradient** (or the baric gradient as it is frequently termed in meteorology).

Pressure is a scalar quantity defined as a force acting on unit area of any surface perpendicular to it. At any point in a fluid, pressure is independent of direction and, at a given depth z, it is equal to the weight of a 'unit' water column of height z. (A unit column is defined as a column of fluid with base area equal to unity.) This hydrostatic law can be written as

$$p = \rho g z \qquad (3.1)$$

where p is the pressure at depth z, ρ is the density of water (assumed to be equal everywhere in the fluid), and $g = 9.8$ m s^{-2} is the acceleration due to gravity (or, simply, gravity).

Equation 3.1 is observed with amazing accuracy in almost all oceanic processes, including, of course, the large-scale movements discussed in this book. This most important dynamic equation is valid for all oceanic circulation models. It relates the pressure and density fields.

Take the Marsigli box again. On one side of the partition, the water had a higher density than on the other side. The pressure at the level of the lower hole was thus greater on that side. Water began to flow in the direction of the lower pressure, where the density was also smaller, and as the volume of water on the other side became larger, it began to spill over through the upper hole. How can one measure the force that pushed the water through the lower hole?

The common practice in fluid dynamics is to refer all forces to unit volume. The basic equations of fluid motion are formulated in terms of the equalities of such forces. Thus, in order to pass from pressure (a force referred to unit area) to the 'specific' force, one should obviously relate it also to a unit of length. Speaking mathematically, one should take a directional derivative of pressure, or determine the rate of pressure increase or decrease in that direction. (Note that a dimension is used here to formulate a physical law. Methods of this kind constitute the theory of **dimensional analysis**.) The resultant rate gives the force acting on a unit volume of the fluid. If the pressure changes in several directions, this rate will have a maximum in one direction. The rate of change corresponding to that direction is referred to as the pressure gradient (or, generally, the gradient of any scalar field) and it is a vector denoted by grad p. Introducing a rectangular co-ordinate system in space, we can obtain components or projections of grad p along the co-ordinate axes.

One can approximate a gradient component – or the gradient itself, given the correct choice of its direction – by dividing the pressure difference Δp at two close enough points by the distance between them:

$$\operatorname{grad} p = \frac{\Delta p}{\Delta l} \approx \frac{dp}{dl}$$

that is, the gradient is equal to the derivative of the pressure along the direction l.

An important point to be made here is that the gradient is positive in the direction of increasing pressure (this follows from the mathematical definition of the gradient as a derivative). The driving force, of course, is always directed oppositely, toward the lower pressure. Therefore, the exact expression for the force of the pressure gradient appears as

$$G = -\operatorname{grad} p$$

As in any other body of water, the pressure in the ocean increases most rapidly along the vertical. Indeed, in the Marsigli box, the horizontal pressure gradient was proportional to the difference between the densities of Black Sea water ($\rho = 1017.0$ kg m^{-3}) and Mediterranean water ($\rho = 1025.0$ kg m^{-3}). Assuming the partition to be 0.01 m thick, and the box 0.20 m deep, the horizontal gradient in it at the time of opening was

$$\operatorname{grad} p = \frac{9.8 \text{ m s}^{-2} \times 4.5 \text{ kg m}^{-3} \times 0.2 \text{ m}}{0.01 \text{ m}}$$

$$= 882.0 \text{ kg m}^{-2}\text{ s}^{-2} = 882 \text{ N m}^{-3}$$

At the same time, the vertical gradient is always equal to

$$g\rho z/z = g\rho$$

for instance, for the Black Sea it is 9966 N m^{-3}, or ten times the horizontal gradient. Moreover, in the ocean, a density difference of 4.5 kg m^{-3} is normally observed at distances of tens of kilometers, so the horizontal pressure gradients are 10^6 times less.

45

Yet the vertical pressure gradient does not affect the horizontal currents – nor the movements in the Marsigli box. This follows from the same basic equation (3.1). Rewritten for the vertical pressure gradient, it is, as has been mentioned.†

$$\partial \rho / \partial z = g\rho$$

This means that for each particle of the fluid the force of the pressure gradient – directed upwards – is balanced by the gravity force of the Earth. In effect, $g\rho$ is exactly that – the force of gravity applied to unit volume.

The hydrostatic law states that in the ocean there is an almost exact balance of forces acting in the vertical direction. It can thus be said that the ocean maintains a state of hydrostatic equilibrium and that the major motions are caused by horizontal pressure gradients. In order to present pressure graphically and compute pressure gradient, two surfaces in the ocean are usually compared – **geopotential surface** and **isobaric surface**.

The first one is the surface connecting equal values of the quantity D defined as

$$D = gz$$

i.e. the geometric depth multiplied by gravity. This quantity expressed the potential energy (per unit mass) acquired by a body – and, therefore, also the work done by it to overcome gravity – when raised through a vertical distance z, and is called the **geopotential** or gravitational potential.‡ In oceanography this quantity is measured in special units called dynamic meters, defined as

$$D = 0.1gz$$

(z is in m and g in m s^{-2}). Since g is close to 10 (9.8 m s^{-2}, on average), the vertical distance in dynamic meters is nearly equal to the geometric distance in meters. Because of this similarity, the quantity D in Equation 3.2 is called **dynamic depth** in oceanography.

† The partial derivative $\partial \rho / \partial z$ is used here to stress that the quantity in question depends on horizontal co-ordinates as well as on the vertical co-ordinate z.

‡ The potential of a vector field (if one exists) is defined as a scalar field whose gradient is everywhere equal to the values of the vector field.

Geopotential surfaces do not change in time and depend on gravity, which changes only slightly from place to place (by only 2–3%), and the local depth – the distance from the undisturbed geoid (Section 2.5). Therefore, these surfaces provide very stable reference levels: actually they are level (horizontal) surfaces because they are perpendicular to the plumb line (or to the vector **g**) at each point.

The second surface – isobaric – need not be introduced in such detail. Isobars are familiar to anyone who has seen a synoptic weather map on a TV screen – they are lines connecting points of equal pressure. Substitute surface for line, and you obtain the definition of an isobaric surface, which, however, may be a little more difficult to visualize.

The graphic representation of the pressure field can be achieved using two different approaches. Meteorologists prefer to select a geopotential surface and project onto a plane the points of intersection of various isobaric surfaces with this reference surface D_n (Fig. 3.3a). Oceanographers would prefer to select an isobaric surface p_n and plot isolines of equal dynamic depth of the surface p_n (Fig. 3.3b). Therefore, in the first case a family of isobars is presented, similar to those drawn on synoptic maps, while the second case displays the topography of an isobaric surface in dynamic depths or, in other words, **dynamic topography**.

Some relations between pressure field, density distribution and dynamics can be drawn from the following simple considerations.

If the ocean had the same density and if its free surface were horizontal everywhere, then – also assuming horizontally homogeneous atmospheric pressure – we would find, at a given depth z, the same pressure at all points in the ocean, and every geopotential in the water would be isobaric. It is not hard to imagine, however, that such an ocean would be motionless.

Now, suppose that, at uniform density, the ocean surface was not horizontal but had elevations, depressions and slopes. Obviously, these highs and lows would be repeated on each of the isobaric surfaces below: at a constant density, the pressure depends only on z, the thickness of the water layer.

In a real ocean, the distribution of pressure depends on the density as well. Besides, the free sea surface is not isobaric, since atmospheric pressure is not constant. As a result, the isobaric surface inside the water have a complex and varied topography. Each surface has its own peculiar

46

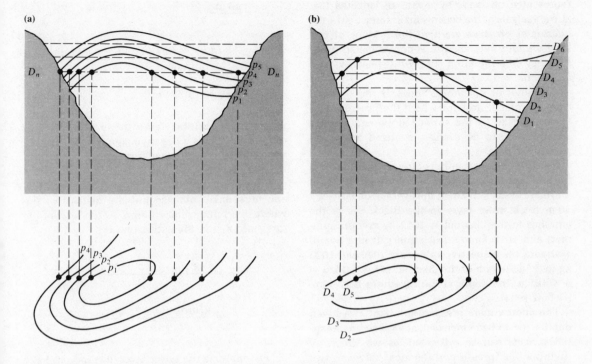

Figure 3.3 Traces of intersections of isobaric and geopotential surfaces in a cross section of the ocean (*top*) and their projections onto a plane (*bottom*). (a) The map of isobars at a given geopotential surface D_n (meteorologic version. (b) The dynamic topography of the isobaric surface p_n (oceanographic version).

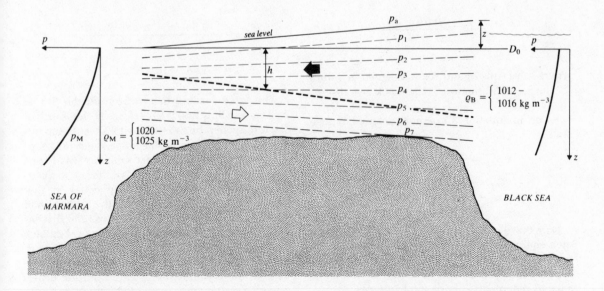

Figure 3.4 Schematic representation of the geometry and pressure field in a longitudinal cross section of the Bosphorus. Surface elevation z, over a geopotential surface D_0, surface slope p_a and slopes of isobars p_1, p_2, ... are grossly exaggerated compared to the slope of the interface h between the waters of the Black Sea and the Sea of Marmara.

shape, and the topography of isobaric surfaces is represented on maps by means of isobaths that allow analysis of the ocean's main force field – the horizontal pressure gradient field. Maps of this kind are routinely used in synoptic meteorology for the determination of wind velocity and direction.

The Strait of Bosphorus presents the simplest case of a planar flow for which the pressure gradients, caused by the surface slope and density differences, can be easily visualized and computed (Fig. 3.4).

Owing to constant influx of fresh water the level in the Black Sea is higher than that in the Sea of Marmara by $\Delta z = 0.26$ m; the average density of a 50 m thick water layer in the Black Sea at the entrance to the channel is 1012 kg m^{-3} (in summer) and 1016 kg m^{-3} (in winter); in the Sea of Marmara, the respective values are 1020 and 1025 kg m^{-3}; the length of the channel is $L = 30$ km $= 3 \times 10^4$ m; the depth at the threshold h is 36 m (Defant 1961).

The approximate pressure gradient at a given depth z (or, to be more exact, at a given geopotential surface) can be estimated, as we know, by dividing the pressure difference between two points by their distance apart. Taking the flow out of the Black Sea as the positive direction, we obtain the gradient force at any given depth z as

$$G_z = \frac{\Delta p}{\Delta l} \frac{p_M - p_B}{L} \qquad (3.3)$$

where subscripts M and B are assigned to values for the Sea of Marmara and the Black Sea, respectively.

By using Equation 3.1, a more detailed expression is

$$G_z = -\frac{g\left[(z\rho)_M - (z\rho)_B\right]}{L} \qquad (3.4)$$

Note that z_M and z_B are not the same on a common equipotential surface: they differ exactly by the value $\Delta z = 0.26$ m. Indeed, all heights, including the surface elevation, are measured with reference to geopotential, horizontal surfaces.

The upper layer in the channel consists almost entirely of Black Sea water, so that Equation 3.4 for it will be

$$G_s = \frac{g\rho_B \Delta z}{L}$$

$$= \frac{9.8 \text{ m s}^{-2} \times 1016 \text{ kg m}^{-3} \times 0.26 \text{ m}}{3.0 \times 10^4 \text{ m}}$$

$$= 8.63 \times 10^{-3} \text{ N m}^2$$

where the subscript s refers to the surface layer (the difference Δz has a negative sign, so the force in the upper layer is directed from the Black Sea into the Sea of Marmara).

When evaluating the pressure in the lower layer, one must take into account the density of the whole overlying water column. At the level of the threshold $z = h$, the expression is

$$G_h = -\frac{g\left[h\rho_M - (h + \Delta z)\rho_B\right]}{L}$$

$$= \frac{g\left[h(\rho_M - \rho_B) - \Delta z\rho_B\right]}{L}$$

Obviously, in the lower layer the gradient force will be defined by the algebraic sum of the 'density' gradient (assuming horizontality of the sea surface) and the gradient caused by the surface slope. The calculations yield

$$G_h = -\frac{9.80\left[36.9 - 0.26 \times 1016\right]}{3 \times 10^5}$$

$$= -1.95 \times 10^{-3} \text{ N m}^{-2}$$

The force causing the water from the Sea of Marmara to flow into the Black Sea is one-fourth of the force generating the opposite current. Accordingly, the discharge of Black Sea water (357 km^3 yr^{-1}) is larger than the influx from the Sea of Marmara (174 km^3 yr^{-1}). The ratio of these two quantities would have been exactly equal to that between the respective forces if the channel had the shape of a straight river bed and the discontinuity surface separating the waters of the Black Sea and the Sea of Marmara had been horizontal. One should bear in mind that all pressure gradients in the Bosphorus originate from the differences in evaporation, precipitation and drainage in the seas linked by the strait or, using oceanographic terminology, from differences in fresh water balance in the seas.

3.3 Coriolis force

We have thus found the first major force acting ubiquitously throughout the ocean. By formulating an equation in which this force is balanced by another force, we will obtain the first mathematical model of the movement of a fluid.

There is a universal force, friction, coupled with any driving force. Actually, if no force is applied, movement will gradually cease, all energy being expended to overcome the resistance of the environment, and ultimately the temperature will rise. In fluid dynamics, this process is known as energy dissipation, but we will return to it later. In the meantime, it should only be noted that outside the zones of direct boundary influence friction in the ocean does not have sufficient strength to balance the horizontal pressure gradients.† In most cases the deflecting force of the Earth's rotation serves as a counterpart for the pressure gradients that arise. Let us discuss this new notion.

Imagine a skater sliding over an ideally smooth, frictionless ice field shaped like a circle rotating at a constant angular velocity ω (Fig. 3.5). In the figure, two marks (A and B) are shown affixed to the circumference of the disk to visualize the rotation. In the absence of friction, the skater will move uniformly and rectilinearly toward an object situated at some distance outside the circle. Observing three consecutive positions of the skater, we notice, however, that he has deviated from mark P on the circle.

A similar experiment connected with the rotation of the Earth is often demonstrated in a planetarium – the Foucault pendulum. The demonstrator pulls the pendulum ball to the edge of the circle and lets it go.

The governing force – a part of the ball's weight – is directed toward the center. The plane in which the pendulum oscillates is seen to rotate in relation to the circle, which has divisions, drawn on the floor. What is the force deflecting the skater and the pendulum?

The common answer would be that the directions in which the bodies move in each case remain unchanged while only the circles under them are rotating. However, in the case of the Earth, this is only half true. It is a fact that the Earth's spinning is responsible for the observed phenomenon – but the plane of the pendulum also

† A steady current, with these forces in equilibrium, is commonly observed in rivers and channels.

(a)

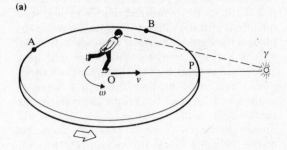

(b)

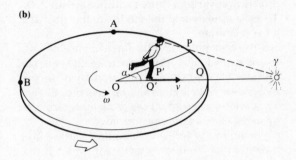

(c)

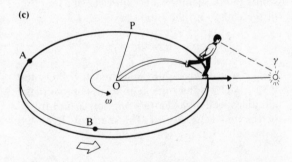

Figure 3.5 Successive phases in the motion of a skater on a frictionless rotating disk as viewed by an outside observer (γ). The rotation of the disk is indicated by the movement of points A and B. The resulting trace is shown in (c).

rotates along with the Earth's surface. The complete solution of this problem is not all that simple, although it can be found in any college text on mechanics. For our purposes, however, an elementary explanation will suffice, which will be obtained by asking our skater to make another racing slide – no easy trick to do in the total absence of friction.

This time we will place our skating-rink right on top of the North Pole, and make it rotate with the Earth. Let us observe the skater from the moment at which he crosses the center of the circle O and starts moving toward Aries (here denoted by γ), the vernal equinox point on the celestial sphere (Fig. 3.5a). It may be recalled that this point has a definite position relative to the fixed stars in the place of the Earth's Equator. Let the skater's speed be $v = 5$ m s^{-1}. After 1 s, the skater will have advanced 5 m along the ray Oγ into the position Q' (Fig. 3.5b). During this time, the Earth turns by an angle α, and the point P that previously lay on the line Oγ shifts to the left. The skater's speed is still directed to the point γ on the celestial sphere, but on the circle it is directed to the point Q. Thus, the skater has shifted to the right from the original direction, covering a certain additional path P'Q'. Therefore, relative to the circle, he traveled with an **acceleration**.

This acceleration and the force connected with it are named after the French mathematician, Coriolis (1792–1843), who formulated several theorems about absolute and relative movements. The Coriolis acceleration can be readily calculated by using elementary physical laws.

For uniformly accelerated motion starting from a state of rest (a uniform movement, which is precisely the case in hand since the velocity v is constant), the acceleration a is expressed as a function of the path l and the time t as

$$a = 2l/t^2$$

It is seen from Figure 3.5b that $l = P'Q'$, but P'Q' $= \alpha OQ'$. The angle α, in turn, is equal to the angular speed of the Earth's rotation ω, multiplied by the time t, i.e. $\alpha = \omega t$. The segment OQ' $= vt$. After substitutions, we find that

$$a = 2\omega v \qquad (3.5)$$

The reader can readily construct a similar scheme for the case of a skater sliding toward the center of the disk. The effect will be the same: the skater will be deflected to the right, given a counterclockwise rotation of the circle, and the acceleration will be expressed by the same relationship, Equation 3.5. A slightly more complex case is a skater moving perpendicularly to the radius rather than along it; here, the centrifugal force has to be taken into account as well. Yet, in this case, too, the Coriolis force has been found to deflect the movement to the right.

Let us now perform some calculations:

$$\omega = \frac{2\pi}{24 \text{ h} \times 3100 \text{ s}} = \frac{6.28}{86\,400} = 0.729 \times 10^{-4} \text{ s}^{-1}$$

Strictly speaking, the denominator should be the sidereal day (86 164 s) but this is irrelevant for our purposes.

It is not difficult to estimate the Coriolis acceleration of the skater. According to Equation 3.5,

$$a = 2 \times 0.729 \times 10^{-4} \text{ s}^{-1} \times 5 \text{ m s}^{-1}$$

$$= 7.29 \times 10^{-4} \text{ m s}^{-2}$$

For more clarity, let us again see how the path of the skater's deviation is determined

$$l = P'Q' = \frac{at^2}{2} = \frac{7.29 \times 10^{-4} \text{ m s}^{-2} \times 1 \text{ s}^2}{2}$$

$$= 3.95 \times 10^{-4} \text{ m} \approx 0.4 \text{ mm}$$

Thus, racing 5 m away from the Pole, the skater will deviate 0.4 mm to the right. After a distance of 5 km, the deviation will be 0.4 m.

This may seem insignificant, but shortly we will see how such seemingly minor effects sometimes control immense natural phenomena.

A few more words about the skater. Note that the foregoing calculations are only 'true' for an observer rotating together with the Earth. From Aries, the skater is seen as coming straight toward the observer. Then, are not all these shifts and accelerations spurious? Indeed, some authors do write about the fictitious and secondary nature of such notions as the Coriolis force, inertia, or friction.

However, the truth is that such forces manifest themselves in an interaction. They are 'fictitious' only insofar as they are considered separately. As soon as the skater strikes against an obstacle, he transmits to it an acceleration according to Newton's third law, and the collision makes the force deflecting the skater a reality. The action of the Coriolis force is not observed in everyday life, because it is too small compared with friction. However, it is only in terms of the deflecting force of the Earth's rotation that one can account for the fact that rivers in the Northern Hemisphere undercut their right (looking along the motion) banks faster than the left ones and that the right track of railroads wears earlier than the left one.

Let us now evaluate the Coriolis force F_C for three cases.

For a skater whose mass is 60 kg moving at the speed $v = 5$ m s^{-1}, we have

$$F_C = ma = 60 \text{ kg} \times 7.29 \times 10^{-4} \text{ m s}^{-2}$$

$$= 4.37 \times 10^{-2} \text{ N}$$

or four-thousandths of the skater's weight.

For the Mississippi River, which carries 19 000 tonnes of water every second at a speed of 0.6 m s^{-1}, the lateral pressure is 1600 N.

For the Kuroshio Current, which carries 65 million tonnes of water per second at a speed of 1 m s^{-2}, we obtain

$$F_C = 2\omega mV = 2 \times 0.729 \times 10^{-4} \text{ s}^{-1} \times 6.5 \times 10^9 \text{ kg}$$

$$= 9.48 \times 10^6 \text{ N}$$

This means that each meter of the current's length feels the impact of a force five times the weight of water flowing per second in the Mississippi – the equivalent of ten heavily loaded freight trains. These numbers are certainly impressive.

Our next step is to find a method for determination of the Coriolis force at any point on the Earth, especially where the surface is no longer perpendicular to the axis of rotation as on the Pole.

To begin with, imagine a rotating circular cylinder bounded by planes perpendicular to its axis. All the foregoing discussion would be valid for the cylinder's end faces. If the cylinder rotates in the same direction as the Earth, the Coriolis acceleration on the upper face ('North Pole') will be directed to the right, and on the lower face ('South Pole') to the left of the direction of motion. What would happen, however, on the cylindrical surface? What is the Coriolis acceleration there? For one thing, it can be readily shown that the relationships given above would no longer hold for it (Fig. 3.6).

Consider movement along the perimeter of the upper cylinder face. The Coriolis force there is directed rightward, that is up or down with respect to the cylinder's lateral surface. This is, in fact, true of the entire cylindrical surface as well: a body moving in the direction of rotation will be pulled off the surface upward while one with the opposite movement will be pressed down toward the interior of the cylinder.

The equatorial belt of the Earth bears a great resemblance to such a cylindrical surface. The Coriolis force here acts vertically to the planet's surface rather than horizontally along it. However, we are interested in the horizontal projection (or component) of the force. We know that at the Pole

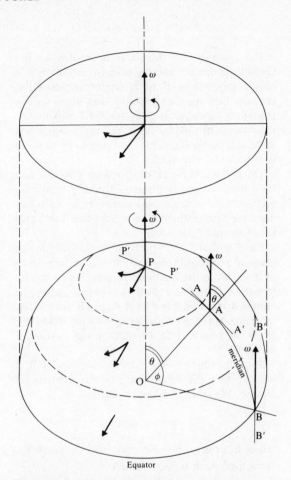

Figure 3.6 Diagram of velocities and accelerations on a rotating Earth illustrating change of the Coriolis parameter with latitude.

it is $2\omega Vm$ and at the Equator it is zero. What are its values between these two extremes?

In Fig. 3.6, the bold arrow represents the angular velocity vector (the conventional notation is to direct the vector so that the rotation appears counterclockwise seen from the vector's end). Comparing the arrows at points P and B, we readily see that the projection of the Coriolis force in the plane of the horizon, such as P'P', A'A', B'B', is a direct function of the vertical component of the angular velocity vector. Indeed, at P the projection onto the vertical is equal to ω, and on the Equator it is zero.

At the point A, the projection onto the vertical is

$$\omega_A = \omega \cos \theta$$

and, since $\cos \theta = \sin \phi$, where ϕ is the latitude of the point A, the complete expression for the horizontal component of the Coriolis acceleration is

$$a_C = 2\omega(\sin \phi)V$$

The quantity $f = 2\omega \sin \phi$ is known as the Coriolis parameter, and it has an important role to play in geophysical fluid dynamics. Incidentally, we now can see that an error was made in the preceding calculation of F_C for the Kuroshio: F_C must be multiplied by $\sin \phi$. For the latitude $\phi = 30°N$, this factor is 0.5, i.e. the Coriolis force there is reduced by one-half.

The balance between the pressure gradient and the Coriolis force is discussed in the following section. In the meantime, however, let us introduce the first mathematical model that has a real oceanographic meaning.

Imagine that a 'fluid particle'† is moving in the ocean at a constant velocity V. It will be deflected – to the right in the Northern Hemisphere – by the force $F_C = fVm$. Can a steady-state movement exist here and, if so, what is it like? To put it differently, what constant force equal to F_C but directed oppositely can be acting on the particle and pushing it to the left?

Since the movement is curvilinear, we will first test whether the force could be the centrifugal force C. We write the force equation

$$C = V^2m/R$$

where R is the radius of curvature of the particle's trajectory. And, if $F_C = C$ then

$$fV = V^2/R$$

From this relationship, R is found as

$$R = \frac{V}{f} = \frac{V}{2\omega \sin \phi}$$

Since the movement is stationary, V is constant and the radius must also be constant. This means that the orbit of the current is circular.

As a matter of fact, currents of this kind are not just theoretically possible but they are frequently observed in Nature.‡ Recently, when examining data on long-term buoy observations in the Black

† A term in fluid dynamics that denotes a small volume, for instance 1 cm³, of a fluid, such that its movement does not differ significantly from that of adjacent such volumes.

‡ The classic inertial 'cyclii' observed by Gustafson and Kullenberg in 1933 in the Baltic Sea are reproduced in most oceanography texts (see, for example, Neumann 1968).

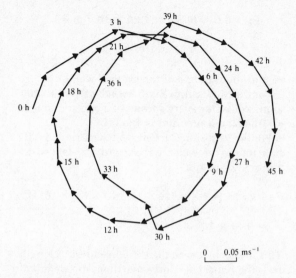

Figure 3.7 Time history (vector diagram) of a current measured in the Black Sea close to the Caucasus coast in 1965.

Sea off the Caucasian shore for 1964–5, this author discovered a cyclic current at a depth of 300 m that existed for more than three days, and had an average velocity of 0.039 m s⁻¹ (Fig. 3.7). We can readily evaluate the radius of its orbit and the turnover time (T) of particles. At the latitude of the Black Sea ($\phi = 42°$),

$$R = 0.039/(2 \times 0.729 \times 10^{-4}\ s^{-1} \times 0.669)$$

$$= 3.99 \times 10^2\ m = 0.4\ km$$

The period is, obviously,

$$T = 2\pi R/V = \pi/(\omega \sin \phi)$$

but $\omega = 2\pi/T_E$, is the period of the Earth's rotation. Hence

$$T = \frac{T_E}{2 \sin \phi} = \frac{24}{2 \times 0.669} = 17.5\ h$$

The period T is sometimes referred to as the **pendulum day**. It is the time taken by the plane of the Foucault pendulum to complete a full rotation cycle at the given latitude. Observations show that the time of complete rotation of the vector (17.4 h) agrees with the calculations perfectly. The existence of such long-lasting rotary currents suggests that frictionless movements are typical at certain depths, and these are important to the understanding of flow mechanisms in the deep ocean.

A current of this kind can certainly take place only after all the initial forces that triggered the movement have ceased to act. A mass of water is then 'released' to keep moving by inertia. For this reason, currents of this kind are called **inertial flows**. Their model – the above expressions and calculations – exemplifies schemes used to explain dynamic phenomena in the ocean. The example also shows a practical application of the modeling procedures shown schematically in Figure 2.16: observations postulating a balance of forces in the ocean; analyzing conditions under which this balance is maintained; calculating, or formulating in qualitative terms, what field of currents is to be induced; and drawing conclusions regarding the model's adequacy as a description of real phenomena. To an extent, this procedure also exemplifies more complex oceanographic models based on the solution of differential equations.

3.4   Geostrophic balance

The following model will help us study the nature of an important variety of stationary motion in the ocean – the **geostrophic current**. This occurs when the horizontal pressure gradient acting on liquid particles is completely balanced by the Coriolis force, while all the other forces, friction and accelerations are absent. Such a balance of forces is referred to as **geostrophic balance**. The interaction of pressure, density (mass) and current fields can be easily visualized by means of a vertical cross section passing through two oceanographic stations A and B (Fig. 3.8a). We assume that the vertical density distribution at these stations has been measured and the pressure is known at every point. These characteristics are depicted by isobars p_1, p_2, p_3 and isopycnals ρ_1, ρ_2, ρ_3, . . . , in the form of lines of intersection of the corresponding surfaces with the diagram plane.†

Consider a typical situation. At station A the sea level is higher and the density down to a certain depth is lower than at station B, as shown by the slope of the isopycnals. The upper isobar coinciding with the sea surface is tilted from A to B (atmospheric pressure is assumed to be constant). It follows from the definition of gradient that the tangent of the angle at which the isobar slopes toward the undisturbed (geopotential) sea surface

$(\tan \beta)$ is proportional to the pressure gradient. Therefore, in the upper sea layer the force of the pressure gradient is directed from A to B. Since $p = \rho g z$, the pressure at station B increases with depth faster than at station A since ρ_B is greater than ρ_A. Therefore the angle of the isobars and tan β decrease with depth, and at a certain level HH the isobar is parallel to the geopotential horizontal surface. Below HH the isobar slope changes direction and increases with depth until ρ_A is equal to ρ_B.

The force of the pressure gradient (vector **G**) can thus be regarded as the sum of two forces: the force **G₁**, due to the sea surface slope, does not change along the vertical, while the other force **G₂**, caused by non-uniformity of the density field, increases down to a depth where the isopycnal surfaces become horizontal. The vectors **G₁** and **G₂** and their sum **G** are shown graphically in Figure 3.8b.

What geostrophic current would correspond to this pressure gradient distribution? The gradient force is balanced by the Coriolis force, which is proportional to the velocity and directed along the normal to the right. The gradient force thus should be directed to the left, also along the normal to the velocity vector. Obviously, the geostrophic current would always be directed so that the higher pressure (at a given level surface) would be on the right. In other words, the isobaric surfaces are tilted to the left if one is looking along the current in the Northern Hemisphere and to the right in the Southern Hemisphere – Figure 3.8c illustrates this point.

It follows that no forces act in the direction of the geostrophic current itself – this is similar to the earlier model of inertial cyclic current. This will always be the case as long as the Coriolis force balances only one force. Naturally, such movements must be in a steady state: there is no force to decrease or increase the velocity.

The velocity of a current can be found from the expression for geostrophic balance

$$G = 2\omega(\sin \phi)\rho V \qquad (3.6)$$

We now calculate the velocity component V_2 due to the gradient G_2. We have

$$G_2 = -\frac{p_B - p_A}{L} = -\frac{D_B - D_A}{\rho L}$$

where $D = gz$ is the dynamic depth of the isobar p. Then by (3.6),

† An isopycnal surface is defined as one connecting points of equal density.

(a)

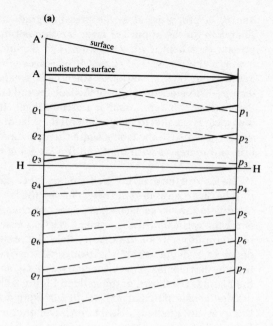

(b)

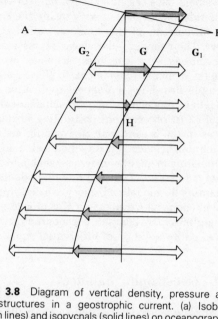

Figure 3.8 Diagram of vertical density, pressure and force structures in a geostrophic current. (a) Isobars (broken lines) and isopycnals (solid lines) on oceanographic cross sections between two stations, A and B. HH is the horizon with zero horizontal pressure gradient. (b) Diagram of composition of the pressure gradients due to surface slope G_1 and inhomogeneous density G_2. (c) Vector diagram of forces and velocities of geostrophic current (oceanographic cross section rotated with respect to the plane of the drawing). The pressure gradient G at all levels is balanced by the Coriolis force F_C, and the geostrophic velocity V is normal to these forces everywhere.

(c)

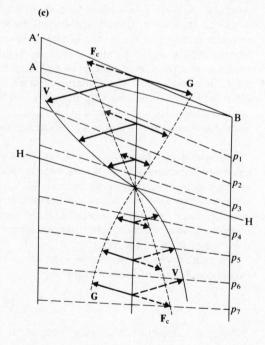

$$V_2 = \frac{D_B - D_A}{2\omega(\sin \phi)L} \qquad (3.7)$$

This is the expression for the relative velocity of a current calculated in relation to the ocean surface. The absolute velocity of a geostrophic current

is obtained by adding to V_2 the velocity V_1 with the corresponding sign determined by the surface slope. In order to do that, one must define the force G_1 as was done for the Bosphorus. However, the topography of the open ocean surface cannot be determined with adequate accuracy, even with the present-day space techniques. Therefore, this method of calculating the absolute velocity of geostrophic currents is difficult to apply to open ocean areas and can only be effectively used in straits where the position of the sea level is known from shore gauges on the banks.

This difficulty can be solved by using the remarkable feature of geostrophic currents already known to the reader – the level where the forces G_1 and G_2 are balanced and the current velocity is zero. By computing the position of this level at many points in the ocean, one can find the surface of zero geostrophic velocity or, as it is usually called, the **depth of no-motion** or 'zero surface'. Its position can be found using some hypotheses for the oceanographic fields, which will be discussed in Section 3.9. The zero surface is a more convenient reference level than the sea surface. The distance between the isobaric surfaces (in dynamic units) is counted up from the zero surface; this characteristic quantity is referred to as the **dynamic height** (d) and its introduction in Equation 3.6 gives the absolute geostrophic velocity above the horizon HH. Below the zero surface the current velocity can be determined by the method used to find V_2.

The two-dimensional structure of the geostrophic current field can be depicted by the dynamic topography of isobaric surfaces in the ocean. We have already described how the current direction is determined in relation to the isobar

slope. The lines of equal dynamic heights are simply streamlines of the geostrophic current, which is directed so that on the right the isobaric surface is higher than on the left and the surface slope (i.e. its gradient) is evidently (see Eq. 3.7) proportional to the current velocity.

Figure 3.9 shows the dynamic topography of the surface of the Atlantic Ocean. Comparing this diagram with Figure 2.2 we find on it nearly all the currents discussed as well as some minor currents. The Gulf Stream with its highest velocities is marked by the greatest surface slope. Hence the rapid determination of surface topography by space altimetry has become so important for large-scale reconnaissance (Section 2.5).

The method of geostrophic calculations – from temperature and salinity measurements to maps of

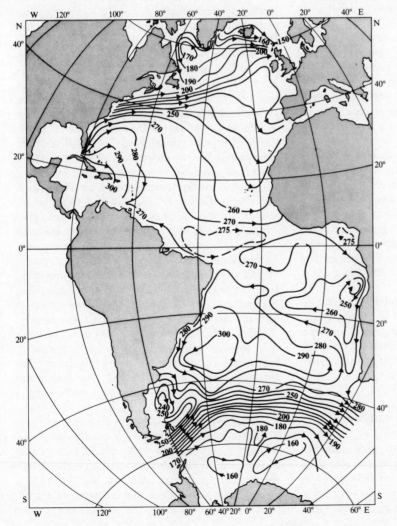

Figure 3.9 Dynamic map (topography of the free surface in dyne cm) of the Atlantic Ocean. (Adapted from Burkov 1980.)

dynamic topography – has become known as the **dynamic method**. It was first used by W. Helland-Hansen and I. Sandstrom in 1903 and was destined to play a major role in oceanographic studies.

Until recently, basic knowledge on general circulation in oceans and seas has been provided by computations using this method. Almost everywhere in the ocean where the dynamic method has been used, computed current fields were largely in agreement with the existing information on oceanic flows. Data on temperature and salinity collected by the oceanographic research vessels 'Michael Sars' and 'Armauer Hansen' (1904–23) and especially 'Meteor' (1925–7) made it possible to calculate currents at various depths all over the Atlantic. The idea of water circulation has changed little for three decades, despite a general improvement in measurement techniques.

In the Northern Hemisphere, all the major gyres with a distinct increase in current velocities near the western coast were found by geostrophic calculations. Only in the very upper layers of the oceans and their equatorial regions, where the geostrophic regime is disrupted (f is close to zero), do the computed velocities differ somewhat from the observed ones. We will discuss these peculiarities later.

We have thus established the remarkable feature of the ocean: from a certain depth, the geostrophic regime corresponds to the actual large-scale flow patterns in the major portion of the World Ocean. Now that we know how the main circulatory system of the ocean operates, we can restate the major problem of dynamic oceanography: which of two factors – differential heat or wind – is primarily responsible for the existence of the geostrophic regime.

We shall now try to apply models of motion to explain the driving mechanisms of ocean circulation. Simple models will be employed to determine the relative importance of each of the factors – evaporation and precipitation, temperature differences and wind.

3.5 Climatic factors and currents

We will first discuss how currents arise under the action of precipitation and evaporation and then discuss the effects of temperature differences in the meridional direction. All these are referred to as **climatic factors**. Assuming the existence of a geostrophic balance, suppose that the ocean has a uniform density. In reality – even disregarding the influence of the non-uniform influx of heat to the ocean – precipitation and evaporation always alter the seawater density, diluting it or making it more salty. In order to study the 'pure' effect on currents of freshwater influx through the ocean surface, we must assume that the entire water body is fresh.

One of the first theoretic models of movement in the ocean – the model of S. S. Hough, later refined by R. Goldsbrough (see Stommel 1957) – was based on this simplification: i.e. in a freshwater ocean covering the Earth as a layer of constant depth, currents are induced only by evaporation and precipitation. The models of Hough and Goldsbrough differed in their patterns of precipitation and evaporation zones.

In the Hough model, precipitation covers the Northern Hemisphere uniformly while evaporation occurs from the surface of the Southern Hemisphere. An excess of water in the north and the resulting bulging of the sea surface create the force of the pressure gradient G_1 directed along the meridians to the south. It is seen from Figure 3.10a that this force is greatest at the Equator, which separates the precipitation and evaporation zones. The force G_1 should be balanced by the Coriolis force (geostrophic balance) so that in the Northern Hemisphere the currents would be deflected to the right from the direction of the gradient force, and to the left in the Southern Hemisphere. The following simple flow-field is to be generated: at all depths in the Northern Hemisphere the geostrophic current will be directed along the circles of latitude from east to west, and in the Southern Hemisphere from west to east. With water constantly supplied to one hemisphere by precipitation and removed from the other by evaporation, the motion in the Hough model will never achieve a steady state. Figure 3.10 displays two moments in the evolution of this model. The change of water level (ocean depth) can be estimated by the 'solid' land area indicated by the broken line (hatched in the imaginary cut-open Earth). The horizontal arrows in Figure 3.10b show the direction of the current.

In the Goldsbrough model, evaporation and precipitation zones are divided by a meridional plane. Besides, their intensity varies within each hemisphere in a more complex pattern than in the Hough model. For this reason, the velocity field of this model is not as simple as that of Hough. It is possible, however, to trace the streamlines in the

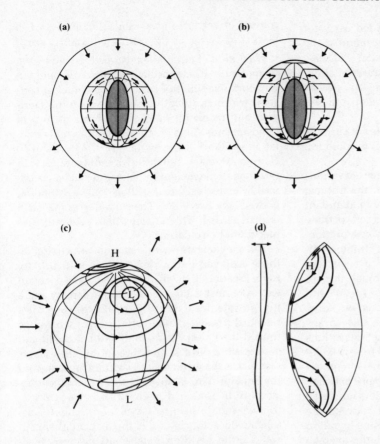

Figure 3.10 Large-scale circulation generated by inhomogeneous evaporation–precipitation processes. (a & b) Two instants in the development of the Hough model; (c) the Goldsbrough model; (d) Stommel's interpretation of the Hough–Goldsbrough model for a meridionally bounded ocean basin.

Goldsbrough model. Initially, the gradient force G_1 is directed along the normal to the meridian, separating the evaporation and precipitation zones. The Coriolis force strives to deflect the current induced by the gradient force away from the Equator: to the right, i.e. to the north, in the Northern Hemisphere and to the south in the Southern Hemisphere. Obviously, this corresponds to a tendency for gyration in each 'quarter' of the Earth (Fig. 3.10c). It would be too complicated for us here to follow in detail the flow patterns calculated by Goldsbrough. Using a sophisticated mathematical technique, he found that two cells of high and low pressure are formed in each hemisphere (marked by letters H and L in Fig. 3.10). Excess of water moves from the precipitation zone to the evaporation zone in the plane of the Equator and thus ensures stationary motion in the model.

The picture of circulation in the models of Hough and Goldsbrough is fairly remote from the real situation. For a long time, therefore, these studies appeared to be very abstract to the majority of oceanographers. It was only in 1957 that Stommel demonstrated that the 'evaporation–precipitation' model of Hough and Goldsbrough could be brought near to reality. This required just one minor change: Stommel introduced into the model meridional boundaries (coasts) while leaving intact the fields of evaporation and precipitation along the meridian – precipitation in the Northern Hemisphere and evaporation in the Southern Hemisphere. The result was remarkable: Hough's and Goldsbrough's gyres lost their symmetry and shifted westward, with streamlines crowding near the coast. Hence a narrow boundary current has been formed that resembles the Gulf Stream! Note that the direction of this current was opposite to that of the real Gulf Stream, because flows are set into motion by a very simple distribution of evaporation and precipitation zones (shown graphically on the left of Fig. 3.10d).

Stommel (1957) showed that in this model it was possible to reproduce all the major gyres of global circulation by introducing more complicated fields of precipitation and evaporation, i.e. adjusting the currents to the real situation by means of an imaginary field of external factors. Hence, ocean currents can be induced by vertical movements:

indeed, precipitation can be regarded as water flow vertically downward and evaporation as the opposite, upward flow. This conclusion is of basic importance for understanding the causes of circulation at greater depths where nearly perfect uniformity prevails and all movement is determined by vertical motions.

We will now estimate the velocities of currents caused by the unequal distribution of evaporation and precipitation zones.

The greatest differences in surface elevations over the undisturbed level occur in the tropical ocean. Here the zone of doldrums with abundant tropical rains (0–10°N) lies near to the zone of trade winds (10–20°N) conducive to heavy evaporation. However, at lower latitudes the geostrophic balance is disrupted. Let us see how currents are affected by flows of fresh water through the sea surface in the latitudinal zones of 10–20° and 20–30°N. From the earlier discussions, we know how to evaluate current velocity with an accuracy sufficient for our purposes. The surface elevations in the meridional direction are found from calculations of the layers annually evaporated or coming as rainfall (Korzun 1974). The difference between the two above zones is $\Delta z = 50$ cm. It is now easy to determine the pressure gradient between the two zones, $\Delta p = \rho g \Delta z$ and then, by using Equation 3.6 for geostrophic balance, obtain the average velocity as

$$V = \frac{g \Delta z}{L 2 \omega \sin \phi}$$

where L is the distance between the zones, equal to 1.1×10^6 km. We find that

$$V = \frac{9.8 \text{ m s}^{-2} \times 0.5 \text{ m}}{1.1 \times 10^6 \text{ m} \times 2 \times 0.729 \times 10^{-4} \text{s}^{-1} \times 0.423}$$

$$= 0.07 \text{ m s}^{-1}$$

Since the level decreases from south to north, this current must be directed from west to east, that is, against the prevailing North Equatorial Current (Fig. 2.2), which has velocities of 50–100 cm s^{-1}. In other areas V is even smaller.

Calculations show that the velocity of currents caused by the unequal distribution of precipitation and evaporation zones averages from 1 to 3 cm s^{-1}. Thus, this driving factor is unlikely to be responsible for the existing current systems. Apparently, there must exist more powerful

sources of vertical flows – and they have in fact been discovered, as the reader will shortly see.

What kind of circulation can be induced by the temperature difference between the Poles and the Equator? Imagine the ocean as a channel of constant width and depth stretched along the meridian and receiving different amounts of heat at opposite ends. This idealization of thermal circulation is found in the models of N. Zubov and V. Hansen (Mamaev 1962) shown on Figure 3.11. In this model, movement in the ocean can be visualized by using just one meridional plane extending between the two Poles. The curvature of the Earth is disregarded. This is sometimes referred to as **meridional circulation**.

The movements in this motion are similar to those observed in the Marsigli box or in a drafty room (Section 3.1). Near the Equator the constant excess of heat leads to lower water density and a higher water level due to thermal expansion. As a result, the water of the surface layer will move toward the Poles. There, they will be cooled and sink, creating excess pressure in the deep layers that makes the water near the bottom move toward the Equator. These movements are no longer geostrophic. In order to describe the balance of forces, friction should also be taken into account. This scheme does not contradict the real circulation of water in the meridional plane, but the current velocities predicted by this model are too small.

The temperature difference above, between high and low latitudes, is thus unable to create the currents observed in Nature. However, evaporation, precipitation and temperature differences acting together are able to induce a meridional circulation in the ocean, which forms a temperature–salinity distribution or a **thermohaline** meridional structure of the ocean. In reality, both deep and surface circulations are responsible for vertical movements and spreading of water in deeper strata. Now, as a first approximation, we will discuss only the effects of climatic factors.

Several processes cause water to sink. The most powerful sinks are located in the polar and subpolar regions. Surface water becomes colder and denser than the water below it, and thus sinks. After freezing begins, some salt is injected into the water, because large and orderly ice crystals cannot hold all the heavy salt molecules. As a result, the dense water descends to a depth at which it is neutrally buoyant or to the sea bed. Since cooling is relatively uniform along high latitudes, a belt of cold water moving equatorward should exist. At

(a) Equator-Pole meridional cross section   **(b) Velocity profile**

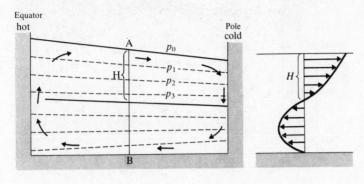

(c) Pole-Pole meridional cross section

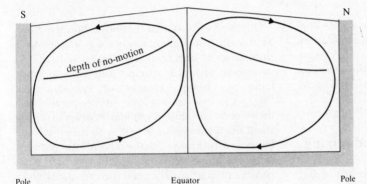

Figure 3.11 (a & b) The Zubov model and (c) the Hansen model of meridional circulation caused by thermal contrast between the Poles and the Equator. The density distribution in the Zubov model shows the reversal of sign in the Pole–Equator pressure gradient (a) and corresponding velocity profile (b). Hansen's model exhibits the depth of no-motion in the meridional section.

the Equator, where water from both hemispheres converges, an upward motion should prevail, but it cannot reach the surface because of the downward motion in the tropical zone.

A high evaporation rate in the belts of the trade winds increases salinity and, hence, density of water, thus inducing convective overturn. Obviously, *density stratification*† should be most pronounced in the tropical zone, because the warm surface waters and cold waters from the polar regions come into close contact. The excess of water beneath the tropical zone causes polarward motion in the intermediate layers. Since the differ-

† The term 'stratification' in general applies to a non-uniform vertical distribution (layering) of characteristics. The ocean has a stable density stratification: density increases with depth. This does not mean that temperature and salinity, which determine the density of sea water, cannot vary in diverse patterns. However, at least one of these characteristics always varies so that density increases downward. In the rare instances where the density stratification is unstable, convection sets in and mixing occurs across a layer.

ences in precipitation and evaporation along the meridian are quite complex (they are shown schematically at the top of Fig. 3.12), several other circulations may evolve at moderate latitudes. The vertical density gradient decreases in a polarward direction. Later (Ch. 7) it will be shown that these moderate-latitude circulations are formed primarily by wind action.

All this circulation which is governed by changes in temperature and salinity along global climatic belts is called **thermohaline circulation**. Since temperature changes play the major role in density stratification, the layer of pronounced temperature decrease (shown by the shaded area in Fig. 3.12) is called the **main thermocline** or the 'permanent thermocline'.

This picture of thermohaline circulation misses many important details. Besides, the vertical motions are strongly affected by surface circulation via the processes of convergence and divergence mentioned in Section 2.3. However, this simple model will be helpful in further dis-

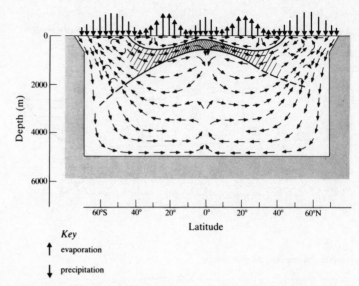

Figure 3.12 The meridional circulation in a rectangular ocean driven by the combined action of evaporation–precipitation processes and heat fluxes across the surface. The shaded area represents the ocean thermocline.

Key

↑ evaporation

↓ precipitation

cussions of surface currents. Later we will be able to modify the picture of deep-ocean circulation, when the mechanisms of wind-driven currents become known to us. The first step in this direction is to study how wind energy is transferred to the ocean depths.

3.6 Turbulence, viscosity and mixing

Any moored sensitive instruments continuously recording the temperature, velocity and direction of currents or any other characteristic of sea water show that these properties are far from being constant in time and space. A veritable host of random and chaotic variations is usually imposed on the regular, lawful changes of any characteristic, thus revealing the existence of turbulence, the phenomenon inherent to most natural flows.

The notion of **turbulence** is applied to every motion of a continuous medium where the orderly, 'average' motion of the large mass is disturbed by random – with respect to the average motion – continuously varying eddies, whirls, oscillations or spurts transferring smaller volumes of substance in every direction. A turbulent environment is permeated by these curvilinear jets and vortices of various sizes. Normally, we ignore them because we are mostly concerned with the average current. Yet, water in rivers, seas, or oceans, as well as flows in many technological processes, are turbulent, and so are the atmospheres of the Sun and Earth, the stellar gas and the constellations, and even the galaxies.

Opposite to turbulent motion is **laminar flow**. This occurs much more rarely – e.g. in fine blood capillaries of our body circulation and in the movement of highly viscid fluids such as vegetable oil, honey, treacle, or volcanic lava (Fig. 3.13). In a laminar flow, jets are parallel and the velocity variation is gradual. In order to understand the mechanisms of energy transfer between layers in a fluid as well as the relaxation of currents caused by the force of friction, it is useful to compare the properties of laminar and turbulent flows.

Imagine two continuous fluid layers moving parallel to each other with different velocities. Because of Brownian motion, a continual exchange of molecules will take place between the two layers. The faster layer will give some of its molecules to the slower one and in turn receive slower molecules. As a result, the fast layer will lose part of its energy to speed up slow molecules of the neighboring layer and will thus sustain a force of friction. This force can be evaluated and compared to the force of friction between two hard surfaces. It can then be assumed – incidentally, Newton was the first to do so – that friction in a laminarly moving fluid is greater when the velocity differential between the layers is large and they are close to each other. Therefore, the force of friction per unit area between the layers of a laminar motion τ_m must be proportional to the velocity gradient in the direction transverse to the flow (n), i.e.

$$\tau_m = \rho\mu\frac{\partial \mathbf{V}}{\partial n} \qquad (3.8)$$

60

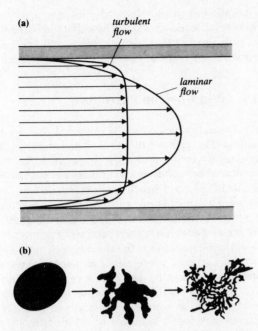

(a)

turbulent
flow

laminar
flow

(b)

Figure 3.13 (a) Transverse velocity profiles for laminar and turbulent flows in a pipe. The turbulent eddies make the average vector field homogeneous in the central part. (b) A drop of dye placed in turbulent flow. In a few instants it will be completely mixed with the surrounding water.

The proportionality coefficient μ is called the **molecular viscosity**. This parameter is specific for each fluid and depends on characteristics such as temperature, pressure, etc.

The transport of molecules, which is responsible for the viscosity or friction in a laminar flow, does not disrupt the overall flow structure. It remains steady throughout the motion, its streamlines being parallel. However, this harmonious picture is disturbed as soon as the current velocity exceeds a certain limit, specific for each particular fluid. The same happens when a laminar flow from a thin tube enters a large reservoir. Eddies of various dimensions are formed and start moving with the main flow, weakening, falling apart and then arising anew. The current becomes turbulent. The transition from a laminar flow to a turbulent flow was first studied by Oswald Reynolds in 1883. Introducing a dyed liquid into the flow, he observed its mixing with the surrounding fluid. As long as the current remained laminar, the dye moved as a distinct thin plume, but after the onset of turbulence, the plume became blurred and soon all the liquid in the tube attained a uniform color.

The Reynolds experiment showed that in a turbulent flow the average motion in the tube is accompanied by transports perpendicular to the tube's axis. Such motions are responsible for the mixing of the dye and for the smoothing of the average velocities in the flow.

To estimate the conditions under which one type of current converts to the other, Reynolds found the dimensionless relationship

$$Re = \frac{Vl}{\mu}$$

where V is the current velocity, l the scale of the flow and μ the molecular viscosity. This ratio, known as the Reynolds number, shows the relative importance of the inertia forces and the molecular viscosity. As the Reynolds number Re indicates, in a fluid of given viscosity the turbulent regime evolves due to either increased velocity or increased scale of flow. Re also explains why more-viscid fluids tend to flow in laminar fashion.

Turbulent phenomena are described by mathematical statistics. The behavior of the particles of a fluid engaged in turbulent motion is as random as Brownian motion. Two particles adjacent at any instant of the motion may be separated at the next moment by an arbitrarily large distance. This makes it senseless to study the behavior of an individual particle. However, it is possible to determine the average characteristics of the motion. In much the same way as with molecular motion, average effects are interesting, such as the transport of various properties of the fluid or impurities by turbulent eddies. Specifically, to understand the mechanism of sea currents, we must find average effects such as the turbulent friction or resistance met by the water. This resistance is determined from the values of the average velocity field, which is done by using the semi-empirical turbulence theories developed in the 1920s by L. Prandtl, J. Taylor and T. Karman.

The basic assumption of these theories is that the complex field of a turbulent current is the sum of two fields (one averaged and reflecting the general transport of water, the other fluctuating and describing the random variation of the velocity vectors or the oscillations arising as a result of turbulent eddies). This method enabled the authors of semi-empirical theories to 'halt' the translation of movement of the fluctuating field and concentrate on the behavior of the fluctuations. An analogy with the kinetic theory of gases was helpful here, in which gas is seen as a collection of an immense number of minuscule balls –

molecules. Analogously to the mean free path of a molecule, i.e. the average distance traveled by a molecule between two collisions, it was assumed that an individual turbulent eddy travels a certain path before losing its individuality and becoming dissolved and mixed up with the surrounding fluid. In analogy with a molecule, this eddy was called a **mole** and its mean free path was called the **length of mixing**.

At sufficiently large Reynolds numbers, the average current generates eddies, traveling in the fluid, transporting the energy of the flow and smoothing the average velocity profile. These effects are reminiscent of those of a laminar current but the transfer of momentum is effected by moles instead of molecules. In analogy with molecular viscosity, the expression for the turbulent current appears as

$$\tau_t = \rho A \, \frac{\partial \mathbf{V}}{\partial n} \qquad (3.9)$$

but unlike μ, here the coefficient of turbulent viscosity, A, is no longer a physical parameter. Not only does it depend on temperature, pressure and salinity, but, before it can be defined, agreement is needed as to what current should be considered 'average'. Velocity fluctuations occur on different timescales – from seconds to days; therefore, the chosen averaging scale should smooth the fluctuations, leaving intact the general motion being studied. This must be the motion whose velocity gradient is used in Equation 3.9.

In other words, the method of computing the turbulent viscosity (friction) should serve to define the average velocity profile. The existing turbulent eddies (their dimensions and intensity) are not themselves described but they merely determine the value of A. In order to stress that A describes momentum transfer by turbulent eddies, A is frequently called the 'eddy viscosity coefficient' or, simply, **eddy viscosity**. If analogous coefficients characterize the transfers of temperature, salinity, or mass, they are called, respectively, **eddy conductivity** and **eddy diffusivity** for salts or mass.

Accurate and sensitive experiments and measurements in a real fluid environment have established that, for any fluid, A is many times greater than μ. Since velocity gradients in a turbulent flow are also usually greater than those in a laminar flow, turbulent friction is virtually the only type observed in the ocean and in the atmosphere.

Now that we are acquainted with the force of friction, the discussion of currents can be resumed. It must be clear by now that the horizontal pressure gradient – whether in a door crack or in a sea strait – can be offset by the force of friction.

3.7 Wind-driven currents

Winds are less fickle than is commonly believed. Earlier (Fig. 1.5) we discovered that during the year calm weather prevails in some areas like the equatorial belts (doldrums) or there are persistent winds in the same direction in other areas like the zones of the trade winds and westerlies. In Section 1.3 it was established that the causes of this stability are related to the general atmospheric circulation. On the sea surface, the wind system produces wind drag, usually referred to as **tangential wind stress**, which sets the upper sea layer into motion. We will now estimate the depth to which the influence of the wind can penetrate. In 1905, W. Ekman developed a model of wind drift for a uniform, infinite and bottomless ocean rotating with the angular velocity of the Earth. Over the surface of this ocean blows wind of a constant force and direction and all movements are attenuated at infinite depth. Since water cannot pile up (the ocean is infinite†), the free surface coincides with the geopotential surface and there must be no horizontal pressure gradient in the water column.

Planar symmetry is clearly expressed in this model: even before solving it, one can be certain that at each depth the velocity vector should not vary from one point to another since the conditions along the horizontal plane would be identical in all directions. This means that vertical variation of the current vector must give a complete picture of the wind drift model. The currents should be attenuated with depth since the tangential wind stress acts at the surface, with a completely stagnant environment assumed at infinite depth.

This model again comprises only two forces: the Coriolis force and the surface force of internal turbulent friction or turbulent viscosity. The former is called a surface force to distinguish it from body forces, such as the gravity force, the Coriolis force and the pressure gradient. These latter forces are applied simultaneously to all points of a small

† Geophysical fluid dynamics uses the notions of infinite and infinity as mathematical idealizations meaning extremely remote from the object in space or time relative to the object's dimensions.

fluid volume, whereas internal friction arises between layers, facets and particles comprising that volume. Here it can be assumed that each particle is affected by two viscous forces – the dragging force from the overlying fast particles and the resisting force from slow particles lying below, i.e. closer to the immobile depths of the bottomless sea. Thus, three forces act on each particle: two viscous forces and the Coriolis force. These interactions produce a much more complex dynamic structure than the inertial and geostrophic models discussed earlier, and it can be traced in the following chain of events.

Wind stress will set the particles of the surface layer into motion, and the Coriolis force will deflect them somewhat to the right from the direction of the wind (in the Northern Hemisphere). Owing to turbulent viscosity, the slab of surface water induces the layer below to move, which deviates to the right even more, but with decreasing velocity. Each layer below is dragged along by the layer above, but deflects to the right, the motion becoming slower with increasing depth. After the process reaches stabilization throughout the water column, the end of the velocity vector will describe a spatial curve known as the Ekman spiral; it is shown in perspective in Figure 3.14. Unfortunately, qualitative analysis cannot determine the angles to which the current vector would be deflected along the vertical. This requires solving a system of equations describing the balance of forces. Meanwhile, the reader will have to take the writer's word that at the surface the current deviates 45° from the wind direction and at a certain depth D actually becomes opposite to the wind at the surface. It has been calculated that the velocity at this depth equals 1/23 of the surface velocity. It is worth noting that the depth D is practically the lowest limit of turbulent transfer because deeper than this the wind-induced currents are negligibly small. Therefore, the level D has been called the **friction depth** and the entire layer filled by the drift currents is called the **Ekman layer**.

By summing the current vectors in the Ekman layer, one finds that the total transport (Ekman transport) is perpendicular to the wind vector (in the Northern Hemisphere). That is, in the Ekman layer as much water is transported in the direction of the wind as in the opposite direction. The vector of the drift current averaged along the vertical, as well as the geostrophic current vector, is perpendicular to the tangential force and is offset, on average, by the Coriolis force.

It is useful to estimate the thickness D of the Ekman layer. It is

$$D = \pi \sqrt{\left(\frac{A}{\omega \sin \phi} \right)}$$

Assuming $A = 100 \, \mathrm{cm^2 \, s^{-1}}$, for the latitude $\phi = 45°$ we have $D = 52$ m; for $\phi = 5°$, $D = 126$ m. As we approach the Equator, D tends to infinity. Hence, the Ekman model would be inapplicable to the equatorial zone.

Measurements of currents far offshore confirmed the existence of the Ekman layer, although a steady-state drift current as predicted by the model (the spiral itself) has never been observed.

Noteworthily, one of the originators of the theory of drift currents was the famous Arctic explorer, F. Nansen. During his voyage on the 'Fram' in the Arctic Ocean, he noticed that drifting ice deviated to the right from the direction of the wind. With remarkable intuition, Nansen explained this phenomenon by the effect of the Coriolis force. He shared his guess with V. Bjerkness, who invited V. W. Ekman, then his student, to find a solution to this problem. After Nansen had explained to Ekman his observations and conjectures concerning the effect of the Coriolis force on the wind-driven currents, Ekman formulated and actually solved the problem of pure drift currents on the same day. This impromptu solution proved to be a basic result, and is still used by oceanographers and meteorologists in theoretic models and practical calculations.

The wind-driven layer, although frequently concealed or distorted by other factors, exists in the atmosphere as well as in the ocean. Wind near the Earth's surface meets with substantial resistance and should also outline the same Ekman spiral.

Though the pure drift model reproduces certain empiric properties of wind-driven current (the deviation of the surface current vector from the wind direction, the depth D to which the wind effects penetrate and the water transport in the Ekman layer), the conditions assumed in this model almost never occur in nature. Wind is practically never uniform over the sea surface. Besides, a wind-driven current may impinge on the coast or a rise of the sea floor, and thus the sea surface always deviates from the geopotential surface. An example of a non-uniform wind field is shown in

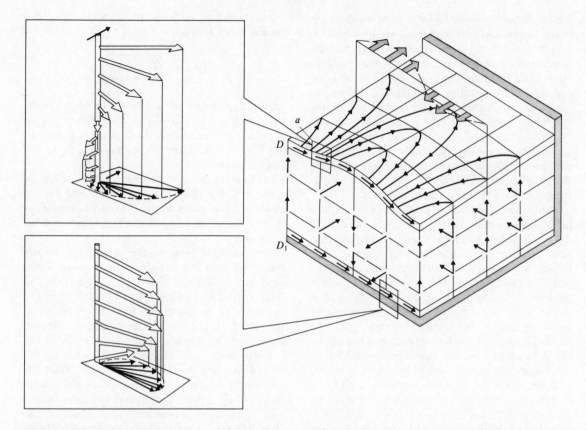

Figure 3.14 (a) Block diagram of motion according to the Ekman model. (After Stommel (1957).) (b) Depth distribution of the flow vector in the upper Ekman layer of thickness D (spiral of drift current vector). (c) Distribution of the current vector in the bottom layer D_1 (inverse spiral). Broad arrows show the distribution of the wind over the ocean's surface; curved lines with arrows are isobaths; black arrows indicate transport of water in the top and bottom Ekman layers.

Figure 3.14 as two wind flows going in different directions (broad arrows). Under each air flow, Ekman transport creates a counterflow in the layer D, so that in the calm belt (the middle of the diagram) the free ocean surface is elevated – the convergence zone. Similarly, divergence zones and depressions in the free surface are formed at the edges of the diagram. An excess of water leads to a downward vertical movement under the convergence zone and an upward movement under the divergence zone. This process of water accumulation or divergence by drift currents is known as the divergence of the Ekman layer, or **Ekman pumping**.

The slope of the free surface creates the pressure gradient G_1, which is the only force driving the entire water column below the D layer. If the gradient G_1 is balanced here by the Coriolis force, the current in the entire water column becomes geo-strophic and is directed along the isobaths of the ocean surface topography. It is only at the ocean floor that the turbulent viscosity becomes significant again, giving rise to an Ekman layer of thickness D_1. The current vector in this layer is attenuated toward the bottom along a spiral bent inversely to that at the surface (Fig. 3.14c). In other words, D and D_1 are not thick (about 100 m) compared to the average ocean depth. The gradient current, therefore, should prevail throughout the water column, which means that below depth D the current at all levels must be almost invariable in direction and velocity. This distribution of gradient and wind currents could exist in a deep, uniform ocean, but no such ocean exists in Nature. Because of stratification, the deep gradient flow below the Ekman layer is concealed by geostrophic currents induced by density differences.

3.8 Baroclinic model

At the beginning of this chapter, we set out to define the main causes of ocean currents by using simple physical models of motion. We established that, theoretically, the geostrophic regime of oceanic gyres may be due to climatic factors. However, this model predicts an implausibly small current velocity, although the density stratification in the meridional plane could be attributed to climatic factors. We found further that wind is the major factor responsible for the high velocities of ocean currents and that its energy can be transmitted to depth by means of turbulence and the pressure gradient caused by surface topography. But the theory of wind currents in a homogeneous ocean predicts a uniform gradient current throughout the water column, and only a thin near-bottom friction layer is responsible for the relaxation of geostrophic flow. No observations have ever confirmed the existence of a huge uniform flow of geostrophic origin sandwiched between two thin friction layers. We are thus faced with the need for co-ordinating the theoretically possible movements of different natures.

In 1936, the eminent geophysicist, Carl Gustav Rossby, offered an original hypothesis. He suggested that the above gradient current is dissipated by horizontal turbulent eddies. Anticipating results somewhat, it should be said that this concept did not solve the problem entirely: horizontal turbulence was essential only near the ocean boundaries and in jet-like currents. Rossby's idea led, however, to a new trend in oceanography – the theory of integral circulation, which will be described in the following two chapters.

Another approach, which proved more fruitful for solving the problem of deep currents, was suggested by the Soviet scientist, P. S. Lineikin, in 1954. But before we describe it, two important concepts, 'barotropicity' and 'baroclinicity', should be introduced.

A fluid is called **barotropic** when the isobaric surfaces are parallel to the isopycnal surfaces, and these in turn may lie parallel to the sea surface. When a barotropic fluid is at rest, the isobaric and isopycnal surfaces are parallel to the geopotential surfaces. If movement is induced only by the surface slope, i.e. the entire body of water moves as a whole under the pressure gradient G_1 (Fig. 3.8b), the barotropic regime is preserved. In the case in which motion results from horizontal density inhomogeneities, i.e. the isobaric and isopycnal surfaces intersect, the barotropic conditions are not satisfied. Such a regime is called **baroclinic**. The pressure gradient G_2 thus signifies the baroclinic component of the geostrophic flow field.

It is obvious that a baroclinic fluid cannot remain motionless, because the intersection of isopycnals and isobars immediately produces a pressure gradient G_2 that sets the water into motion. Another corollary is obvious from the above definitions and considerations: an ocean of constant density is always barotropic, since the notion 'isopycnal' is immaterial in this case.

In view of this discussion, the Hough–Goldsbrough and Ekman models fall into the category of barotropic models. All the models of Section 3.5 associated with the meridional motions are baroclinic. The Lineikin model that we are going to discuss is also baroclinic, because it deals with density inhomogeneities below the friction layer due to effects of Ekman pumping.

Let us discuss the Lineikin model in its simplest form. Similarly to the Ekman model, the ocean is assumed to be shoreless and bottomless, with the wind field over its surface assumed known. At infinite depth, the velocity components – horizontal and vertical – become negligible; the horizontal pressure gradient also disappears. An important property of this model is that it is time-variable, unlike the previous ones. The wind field is 'triggered' at the first instant and then remains constant. It is also assumed that, before the motion starts, vertical stratification of temperature and salinity has been created by transfer of heat and moisture across the surface as in Figure 3.12.

When, after an interval of time, a stationary motion is established in the upper layer, the current structure will be similar to that described by the Ekman model, assuming a homogeneous density of the upper layer. The velocity vector varies here with depth along a spiral curve, the drift transport forms zones of divergence and convergence in the Ekman layer (Ekman pumping) and the ocean surface deviates from the geopotential surface, creating barotropic gradient G_1.

Peculiarities begin beneath the Ekman layer (Fig. 3.15). As we know, viscous forces do not penetrate so deep. Therefore, turbulent friction can be disregarded, the pressure gradients are balanced by the Coriolis force, and the balance of forces is geostrophic. Now consider the vertical currents at the lower boundary of the layer D. They lead to the ascent of dense water under the con-

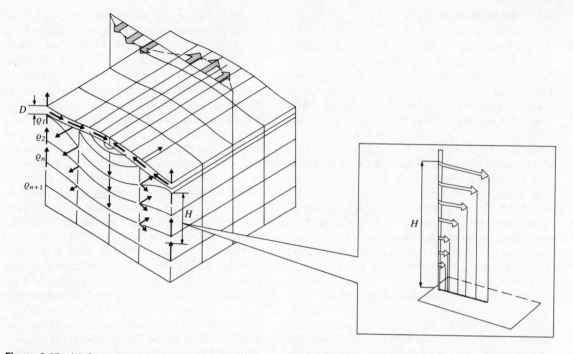

Figure 3.15 (a) Stommel's (1957) interpretation of the baroclinic model of Lineikin. Coast and bottom are absent. Isopycnal surfaces $\rho_1, \rho_2 \ldots, \rho_n$ descend in the convergence zone and rise in the divergence zone of the Ekman layer. At a certain depth the pressure gradients due to the surface slope and inhomogeneous density cancel each other and the motion disappears. (b) Depth distribution of the current vector in the layer of gradient flow.

vergence zone and the descent of light water under the divergence zone, because before the wind started to blow the water had been stratified. As a result, the isopycnal surfaces will not be horizontal, and the fluid below the layer D becomes baroclinic. The gradient force G_2 is thus created at depth. This force counteracts the barotropic pressure gradient G_1. The gradient current gradually diminishes as one moves vertically downward and does not penetrate to great depths. The vector G_2 increases with distance from the surface. At a certain depth H, it becomes equal to the constant vector G_1. This depth, known as the **baroclinic depth**, or **Lineikin depth**, determines the depth in the ocean to which the gradient currents can penetrate. What happens beneath it? Basically, there should be no movement below. However, we know that in a real ocean currents do penetrate to the very bottom, for example, due to sinking at the Poles and rising near the Equator (Fig. 3.12). Therefore, the Lineikin model, in this form, best fits deep movements in inland or closed seas, such as the Caspian Sea or the Black Sea, etc., where convection does not reach the bottom. A modification of this model to describe oceanic conditions will be discussed later.

In the Lineikin model, the departure of the density from its initial distribution at rest is unknown, and is determined together with the current field. The momentum equations, therefore, complement those known in fluid dynamics as turbulent diffusion equations. They are also frequently called **mass transport equations** because they describe the fluxes of a substance in a fixed volume resulting from two processes: diffusion due to turbulence and advection or pure translation of matter by currents.

At constant pressure, seawater density, of course, depends on temperature and salinity. Therefore, some equations are needed to link the easily measured temperature and salinity with the current velocity. Both temperature and salinity transport equations are suitable to this end but Lineikin simplified the problem, reducing the number of unknowns. He showed that water density can, with sufficient accuracy, be assumed to be a linear function of temperature and salinity. This allowed him to formulate a single equation for mass (density) transport – the turbulent diffusion equation – which closed the set of equations for the model of currents in the baroclinic layer.

The Lineikin model clarified why the geostrophic balance prevails in most of the ocean. We now know that the prevailing wind creates not only drift currents but also deep ocean currents. It will be recalled that, in the Hough–Goldsbrough model, evaporation and precipitation appeared to be insufficient to generate vertical currents capable of inducing the horizontal velocities observed in nature. In a real ocean, the vertical movements are generated by the wind as a result of Ekman pumping, and they are responsible for the horizontal non-uniformity of the density field in the body of water. This is the first condition for the existence of a geostrophic current. The other condition, the absence of turbulent friction in the baroclinic layer, is provided by the fact that viscous stresses are essential only near the surface, in the thin Ekman layer. As a result, geostrophic balance prevails throughout the column of ocean water.

Before the baroclinic ocean theory was created, various views were proposed as to the ability of the dynamic method to simulate a real current. In particular, N. N. Zubov and his school believed it to be suitable for calculating any stationary current. Since currents achieve steady state with time, they argued, the density field must also be restructured correspondingly, and we can register this response in the body of water by measuring the temperature and salinity, i.e. the variables required to compute the pressure gradient. Zubov was an ardent proponent of mass-scale geostrophic calculations. He developed a convenient procedure, a special 'numeric graphic algorithm', for quick computation of the dynamic depth. The dynamic method became extremely popular and was acclaimed as a universal method for calculating steady-state currents, at least in Russian and German schools of oceanography.

Another outstanding oceanographer, H. Sverdrup, argued that the dynamic method had its limitations. Since geostrophic currents are determined from deformation of the density field, the currents calculated with this method would be 'density currents', thus neglecting any effects on the free surface slope.

We now know that the dynamic method is suitable for calculating all steady-state currents due to a horizontal pressure gradient of whatever origin if the free surface slope is known, or if the velocity is known at one level. The method can also account for currents caused by non-uniform atmospheric pressure developed by steady-state cyclones and anticyclones, if the atmospheric pressure gradient

manifests itself in a dynamic depth distribution. It is only in the layers where the turbulent friction becomes significant, such as drift currents and the near-bottom friction layer, that the dynamic method fails to reproduce current systems.

3.9   Depth of no-motion

The accuracy of any calculation of absolute geostrophic velocities depends on the position of the 'zero surface', or **depth of no-motion**, the reference level for measuring dynamic heights. Finding this horizontal isobaric surface is therefore an important step in dynamic calculations.

The theoretical diagram in Figure 3.8 allowed us to do it easily. How can it be accomplished in a real situation? The principal difficulty is that the topography of the free surface in the open ocean is largely unknown. Sometimes oceanographers lower the 'zero surface' as deep as possible, believing that currents in the ocean's abyss are negligibly small. However, below the zero surface the pressure gradient in the ocean may be of opposite sign to the surface slope. Therefore, sinking the reference level too deep may lead to error. Additional considerations are thus needed to identify the depth H from data on the vertical density structure. The problem is made easier by the fact that in reality the 'zero' surface is not a real material surface but a certain layer in which geostrophic currents virtually disappear.

The 'zero' surface remained one of the major problems for a long time and was investigated by many leading oceanographers. They were not concerned solely with finding a reliable method of determining the depth H for geostrophic calculations. The position and properties of the 'no-motion' layer as such are important characteristics of the dynamic structure of the ocean.

The depth H can be calculated by a dozen different methods. Most of them assume that, near and beneath the depth of no-motion, the horizontal differences of all characteristics, such as gradients of temperature, salinity, density, oxygen content, plankton, etc., should disappear. The method suggested by A. Defant seemed a most convenient interpretation of this approach. He suggested plotting the vertical profiles of dynamic depth differences for pairs of oceanographic stations. The layer where these differences are constant is assumed to be the no-motion layer. This can be readily understood by visualizing the variation of dynamic depth at neighboring verticals between parallel and horizontal isobars. The map of the depth of no-motion of the Atlantic Ocean, based on numerous observations, agreed so well with the known characteristics of currents that Defant acquired a great many followers. His method was used to map the depth of no-motion in the Black Sea by G. Neumann in 1942, in the North Pacific by A. D. Dobrovolsky in 1949, and in other regions.

Another method makes use of the dynamic hypothesis of the formation of a layer of minimum oxygen concentration. In a nutshell, the lowered oxygen content at intermediate depths observed in all three oceans is assumed to be the result of water stagnation. The vertical turbulent flux of oxygen enriches only the surface layer of the ocean. Deep layers are supplied with oxygen due to the slow advection of cold water from higher latitudes toward the Tropics (Fig. 3.10). For intermediate depths, however, there is no obvious source of oxygen (the Zubov and Hansen model can be recalled here). Following this hypothesis, the depth H can be identified with the layer where oxygen depletion is observed.

Identifying the depth of no-motion is thus reduced to the determination of the total oxygen concentration. However, there are serious objections to these hypotheses, in the light of present-day knowledge of the complexity of oxygen consumption in the ocean.

We will not dwell here on those methods that never became widespread. (In 1962, this writer (Tolmazin 1962) also proposed a method for computing the depth H, which was verified in the South Pacific but failed to acquire a following.) Yet one method utilizing the dynamics of the deep ocean layers deserves attention. It was proposed by O. I. Mamaev (1962).

Before explaining his idea, a contradiction in the preceding discussion that a perspicacious reader might have noticed already should be pointed out. Notably, we established that the geostrophic regime in the ocean is formed without any contribution from turbulent viscosity, except in the thin upper layer, over shallows and in the near-bottom layer, where this force is well developed. This, however, contradicts the statement that all motion in the ocean is turbulent.

Everything falls into place when the magnitudes of the forces are compared. Turbulence and viscosity are small compared with pressure gradient throughout the main body of the ocean, but they become significant at the boundaries. They do exist, however, and should not be disregarded completely. It will be more correct to call the regime prevailing outside the viscous layer *quasi*-geostrophic (i.e. 'nearly geostrophic'). Mamaev took this small correction into account in his study of the depth distribution of turbulent energy.

Turbulent eddies, as we know, carry water substances across the flow, as well as make the flow viscous. If the time-constant density stratification of water becomes less pronounced with increasing depth, it can be reasonably assumed that turbulent energy, which is considerable in the top layers, gradually diminishes toward the bottom. This effect can be easily understood: in a baroclinic ocean, the vertical velocity gradient decreases with depth and the current velocity also decreases. Once a minimum turbulence layer exists and can be determined from some indirect considerations, it can be regarded as the no-motion layer. Mamaev attempted to find a mathematical expression for the depth of the minimum energy of vertical turbulence. In order to relate H to measurable characteristics, the eddy viscosity and eddy coefficients for mass must be known. Unfortunately, these variables have been extremely poorly investigated for oceanic conditions. As a result, Mamaev only succeeded in determining the depth of no-motion for annual climatic conditions over the ocean. (He must be credited, though, for a comprehensive investigation of the problem in general.)

The map of the depth of no-motion in the World Ocean compiled by Mamaev (Fig. 3.16) gives an indication of the depth to which currents penetrate in different regions. It comes closest to the surface in the equatorial zone, where its depth is sometimes about 400–500 m. Toward the Poles, H is lowered, faster in the Southern Hemisphere (to a maximum of 2.2–2.5 km) and slower in the Northern Hemisphere (to 1.8–2.0 km). Major peculiarities are introduced by the boundary currents of the Gulf Stream and Kuroshio. For instance, under the Gulf Stream the no-motion layer across the current becomes almost 1 km deep. However, beneath the Kuroshio the transverse slope of the surface H is less pronounced, perhaps because of insufficient data on temperature and salinity available at that time.

Modern oceanographers still use dynamic methods to determine average flows in the oceans. However, the search for precise information on the depths of no-motion has all but been abandoned. The geostrophic currents are usually computed

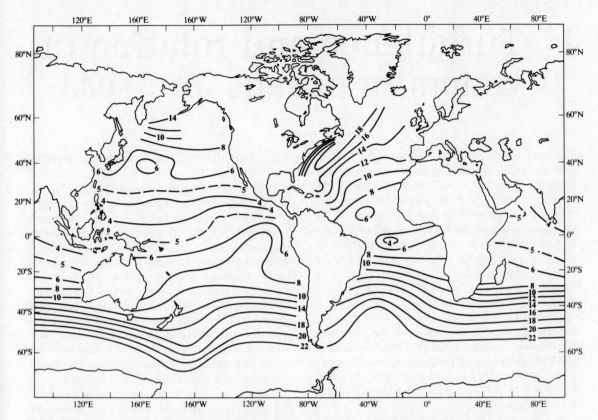

Figure 3.16 A map of the depth of no-motion (hundred meters) in the World Ocean. (Adapted from Mamaev (1962).)

relative to a fixed, rather deep, reference level. The resulting general pictures appear to be sufficient in many applications. However, the depth of no-motion has great educational value since by itself it displays the dynamic structure of the ocean.

To sum up, we can now speak with more certainty about the causes of ocean currents. Wind is responsible for the high current velocities. It gives rise not only to surface drift currents but to deep currents as well. Wind energy is stored in the ocean by restructuring the temperature and salinity fields and also as a result of free surface deformations. These effects induce pressure gradients inside the water column – the force maintaining currents to great depths. Climatic factors, however, are also important elements in the general circulation. The constant and steady action of heating and cooling, precipitation and evaporation, and melting and freezing of ice contribute greatly to meridional density stratification in the ocean, making the entire water body receptive to the wind, and forcing it to behave differently than it would otherwise.

4 Turbulence and rotation of waters in oceans and seas

The models described in the preceding chapter provide useful information on the nature of currents and the density structure of sea water. They do not, however, bring us any closer to understanding the gyration of ocean currents in the horizontal plane. Indeed, gyres of different scale are a characteristic trait of oceanic circulation. As we know, both Pacific and Atlantic waters move along an almost closed trajectory in at least five places; in the Indian Ocean gyres are observed in three or four places, depending on the season. If we could plot on the map of currents in Figure 2.2 the circulations in the seas of the Northern Hemisphere—Mediterranean Sea, Baltic Sea, Sea of Okhotsk and others – we would see cyclonic gyrations of water there as well. Sea gyres, although less steady than their oceanic counterparts, occur almost everywhere.

We also know that eddies appear and vanish in the ocean due to diverse, and often obscure, factors. The structure of oceanic eddies is one of the most exciting and extensively investigated problems of dynamic oceanography. Costly experiments such as MODE and POLYMODE (Section 2.5) have been staged in order to decipher what appears to be a hierarchy of eddies, as will be shown in the next chapter.

We shall attempt to do the same with more modest means – the basic knowledge outlined in the preceding chapters plus certain concepts and theories developed during the past two or three decades. The sequence of presentation will be of great importance here.

As before, we will begin by trying to find a system in an apparently chaotic tangle of events. With eddies it will be convenient to proceed from the 'lowest' pattern of any current system or smallest eddy-like fluctuations that we have earlier called turbulence.

4.1 Turbulence: methods and concepts

Turbulence and turbulent friction were defined in Section 3.6 using an example of a simple flow structure: a planar current with transverse shear. Transition to the three-dimensional current requires adding to the four unknowns – the pressure and three velocity components – another six terms similar to the force τ_t (Eq. 3.9). As in the shear flow, these quantities describe the momentum fluxes due to turbulent velocity oscillations, and are known as **Reynolds stresses** – since turbulent oscillations produce resistance to the main flow. It is thus difficult to construct a mathematical model of a turbulent current since the number of unknowns exceeds the number of equations.

How can a system of equations be closed to make the problem of a turbulent current mathematically solvable? Following the principles set forth earlier in the section on models (see Section 2.7), Reynolds stresses should be expressed in terms of the known mean quantities (velocity and density), as we have done for the viscosity coefficient, proceeding from the nature of turbulence. However, turbulence happens to be a phenomenon for which it is extremely difficult, if not unfeasible, to construct a general theory. In a turbulent flow the velocity field is subject to random variations. While depending on the internal structure of the flow, it also affects that structure itself and responds in an unpredictable manner to external factors. Modern physics sees turbulence as one of the biggest unsolved mysteries of science, and the eminent physicist, Richard P. Feynman, has called it 'Problem No. 1' of present-day physics.

Analytical methods of fluid dynamics have so far been powerless to investigate the behavior of individual turbulent eddies by means of equations. Only recently, with the advent of modern computers, has it become possible to simulate some turbulent flows in two or even three dimensions, and to trace the development of individual turbulent patterns over long periods of time (for instance, see the paper of P. Rhines in *Oceanus* (1976)). However, for many theoretical purposes and practical applications, there is no need to have instantaneous pictures of turbulent flows. What are really needed are the average effects of internal motions, such as the laws of resistance and average transport of substances. Turbulence is therefore investigated with a special statistical tool – the theory of random functions, which originated as recently as 30 or 35 years ago.

The 'failure' of turbulence theory to produce easily observable analytical methods that relate oscillatory and average flow fields did not prevent practical uses of turbulent currents. Unabashed by the actual complications, engineers, of course, keep finding more and more ingenious solutions to problems of turbulent flows in oil and gas pipelines, giant plants for uranium isotope separation by gas diffusion, power plants using the heat from atomic reactors and magneto-hydrodynamic generators, new chemical processes, and so forth.

All these processes are associated with turbulence problems, which are solved by means of the semi-empiric hypotheses described earlier (Section 3.6). Basically, these are relationships between the physical characteristics of the turbulent environment, which are derived from qualitative considerations, but cannot be proven rigorously and require experimental verification. The reader will recall how we made use of the semi-empiric theory of turbulence developed by Prandtl and Karman. We never went further than the notion of the length of mixing and Equation 3.9. For the models we formulated in Chapter 3, the eddy viscosity A was assumed to be a known constant. That was quite sufficient, for our objective was simply to understand the mechanism of large-scale currents, and any detailed study of turbulence would have detracted us from that goal. Now that we know more about the ocean, there are more reasons, and necessity, for returning to the discussion of turbulence. As usual, we begin with a general approach.

The two fundamental methods in the present-day theory of turbulence are dimensional analysis and the principles of symmetry. Along with the possibility of constructing additional relationships to define the characteristics of turbulence, these two methods provide insight into the phenomenon itself.

Dimensional analysis will be discussed later. As to symmetry, we have already used it and will have many more occasions to do so. Specifically, the Ekman model of net drift is an instance of two-dimensional homogeneity or two-dimensional symmetry. Obviously, in this model, the characteristics of a turbulent flow at a fixed depth do not vary from one point to another. The equations of motion for a two-dimensional case of symmetry are greatly simplified. Subsequently, we will make frequent use of symmetry in studying the nature of turbulence in the ocean. The reader will have to learn how to discern the plane of symmetry in each particular approach, which is in fact straightforward.

4.2 Turbulence and stratification

The description of turbulent movements in the ocean can be conveniently started with notions already familiar to the reader. We know that ocean water has a stable stratification, that is, the water density ρ steadily increases with depth. In addition to the main thermocline, there are also seasonal and daily variations of temperature and salinity, which are particularly noticeable in middle latitudes. The layer of a sharp density increase in the downward direction is sometimes called the **discontinuity layer**.

It can be easily proved that vertical turbulence in a stably stratified sea is weaker than in a uniform sea. Imagine a water particle driven downward by a turbulent oscillation. At the end of this movement, the particle will be surrounded by water of density greater than its own. It is thus immediately affected by a buoyancy force equal to the weight of surrounding water that would have occupied the volume containing the particle. This force will be pushing the 'intruder' back to its previous level (Fig. 4.1a). A similar effect must be expected with a particle displaced upward. In this case the buoyancy will force the particle downward. A turbulent eddy is thus affected by forces inhibiting its free vertical movement inside the water column. Only a vigorous eddy of high kinetic energy would be able to counteract the buoyancy force. These considerations suggest that vertical turbulence depends largely upon the balance between the kinetic energy of turbulent motion and the potential energy inherent in the stably stratified water reacting to movement by buoyancy forces. It is naturally assumed that the kinetic energy of a particle is proportional to its density ρ, and the vertical velocity gradient $\partial v/\partial z$. The buoyancy force

(a)

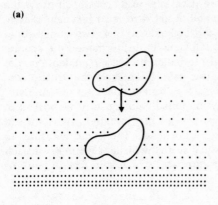

(b)

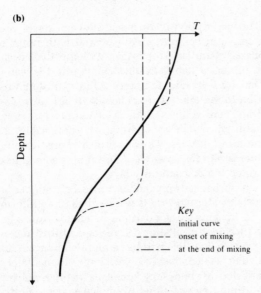

Key

——— initial curve

- - - - - onset of mixing

—·—·— at the end of mixing

Figure 4.1 (a) In a stably stratified sea, a water particle transferred to a level of different density experiences a buoyancy force driving it to its former position. (Density increases with depth in proportion to the density of points.) (b) The evolution of a temperature (T) vs depth curve due to turbulent mixing: at the end of mixing, the discontinuity layer becomes very sharp.

apparently depends on the vertical density gradient $\partial\rho/\partial z$, and the acceleration of gravity g.

We will now attempt to construct a dimensionless relationship between these two products. We write the dimension of each parameter

$$[\rho] = ML^{-3} \qquad [\partial v/\partial z] = T^{-1}$$

$$[\partial\rho/\partial z] = ML^{-4} \qquad [g] = LT^{-2}$$

where M is the dimension of mass, L the dimen-

sion of length and T the dimension of time. It is readily seen that a dimensionless relationship is obtained if

$$Ri = \frac{g \; \partial\rho/\partial z}{\rho \; (\partial v/\partial z)^2} \qquad (4.1)$$

where Ri is known as the Richardson number. This important parameter characterizes the intensity of turbulence, which depends on the stratification and the vertical shear in the current field. From Equation 4.1 it follows that, with unstable stratificaton, $\partial\rho/\partial z < 0$, i.e. the density in upper layers is greater than in lower layers (the z axis is directed vertically upward), then $Ri < 0$. This happens in the case of intense turbulence. At $Ri > 0$, the density stratification tends to attenuate turbulence. Experiments show that, in the oceanic thermocline layer, Ri approximately equals unity, and vertical turbulence all but disappears. A critical value of Ri can be found that represents the boundary between the turbulent and laminar regimes. Specifically, it has been found that at $Ri > 1/4$, virtually no turbulence can be induced by velocity gradients existing in the ocean. The issue, however, needs further clarification.

Where is vertical turbulence most intense? We now know enough about turbulence to be able to answer this question. Since the main thermocline is an obstacle to turbulent eddies, because of high values of Ri there, intense turbulence can only develop below and above that layer. Yet, below the thermocline, vertical density variations are hardly noticeable, and therefore $\partial\rho/\partial z$ is small. At the same depth, $\partial v/\partial z$ is usually small, because only gradient currents, caused by the barotropic pressure gradient G_1, exist (see discussion in Section 3.8). Therefore, at great depth, Ri is the ratio of two small quantities, and so Ri may be quite large at such depths. In physical terms this means that both the kinetic energy of eddies and the potential energy in deep basins are small. If, at great depths, the gradient flow does not change along the vertical, which is the case except for some boundary areas, Ri is high and turbulence is nearly inhibited there.

Vertical turbulence is always present only in the upper ocean layer, where, according to the Ekman model, the vertical velocity gradient is high and water is being constantly mixed by waves and convection. This layer, however, is less than 100 m thick. Besides, intermediate density discontinuity layers may also inhibit vertical velocity fluctuation. Intense vertical turbulence mixes water in the upper layers but enhances the vertical temperature contrasts at the lower boundary of the mixed layers – which ultimately extinguishes turbulence in the water body below. These effects are illustrated in Figure 4.1b.

The intensity of vertical turbulence is thus largely limited by inhomogeneities in the density structure of a water body. What about horizontal turbulence? Apparently, there are no limitations on horizontal eddies.

We are reminded here of the major horizontal eddies (Fig. 2.11) sometimes observed with the aid of infra-red imagery. We will see that in the course of oceanographic experiments, POLYGON-70, MODE-73 and POLYMODE, huge apparently random eddies were documented in various regions of the Atlantic Ocean. Although some of these eddies are a hundred kilometers in size, and can survive as dynamic entities for many months, these patterns fit the definition of a turbulent fluctuation. Obviously, the conditions of horizontal and vertical turbulence in the ocean are quite different.

When a phenomenon tends to evolve with greater intensity in certain directions than in others, it is called **anisotropic**. Anisotropic turbulence means that the slightest instability may cause turbulent eddies to form in certain planes. In the ocean, where the Reynolds number $Re = Vl/\mu$ is always large, because of large l (≈ 1000 km), the conditions for turbulence are always present; a turbulent eddy easily forms in the horizontal plane whereas the vertical direction is 'forbidden' to it.

The opposite of anisotropic is **isotropic** movement – one for which all directions are equally important, such as the flow in a round pipe. We will see later that isotropic turbulence is sometimes also observed in the ocean, although it is different from that in a pipe.

The next important step is to understand the mechanism of interaction of horizontal and vertical turbulence in the ocean. Actually, two things must be clarified: (1) the mechanism by which the velocity field becomes fluctuational, and (2) the external factors causing instability of large-scale ocean currents. This analysis will allow us to determine turbulence parameters, such as the eddy viscosity coefficient, which until now has been assumed to be known and constant.

Before we proceed with the analysis, a few words must be said about one of the major discoveries of modern science – the theory of local isotropy. In contrast to all other largely empiric studies of turbulence, this theory was created by its author, A. Kolmogorov (1941), literally at the tip of his pen, grown on 'the substrate of pure theoretical thinking', to use the phrase of a well known geophysicist.

4.3 Locally isotropic turbulence

Strictly speaking, to say that turbulent motion is absolutely random is not quite correct. There is some order in this apparently random field of motion. The transfer of energy occurs mainly from eddies of larger scale to those of smaller scale. As everywhere else in nature, a large structure tends to break into smaller units similar to that structure.

Kolmogorov showed that the largest eddy is caused by instabilities of the averaged flow and is commensurate with the spatial scale of this flow. Like the average flow from which it takes energy, this largest eddy lacks stability, which makes it break into a number of smaller eddies; these, in their turn, divide into smaller eddies, and so on (Fig. 4.2). Under this fractionation scheme, the energy supplied to the oscillatory field cannot return to the average flow, but will be transmitted down the cascade of smaller and smaller eddies and will ultimately set into motion individual molecules, raising the temperature of the fluid.

One feature of this cascade of energy transfer is particularly remarkable. It is the change of orientation of turbulent eddies in an originally anisotropic flow. This effect can be exemplified using, as before, a horizontal turbulent flow with transverse velocity shear (Fig. 4.2). It is obvious that the largest eddies will retain anisotropy, i.e. they are largely horizontal. This property will be preserved by smaller and smaller eddies in the cascaded process. It can be assumed, however, that, because of the chaotic nature of the process, starting from a certain length scale Λ, the eddies will lose their distinct adherence to the original plane. From that

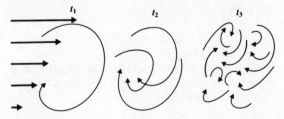

Figure 4.2   Cascaded decay of eddies in a shear flow field.

point on the process may be regarded as isotropic, so that from that size and smaller, no predominant direction of eddies will be observed. It is thus possible in any turbulent flow to determine a certain range of scales which will be locally isotropic. Its upper limit will be Λ and its lower limit λ_m, the scale of molecular interactions.

The **interval of local isotropy** is particularly interesting. At a later stage in the discussion, analysis of its nature will help us to understand the mechanism of energy transfer in the ocean by motions of various scales. For the time being, however, we will take a closer look at the flow of energy of oscillatory motion down the cascade of eddies in the range Λ to λ_m. All the way along this cascade – down to the point at which energy begins to be spent to increase heat – the total oscillatory energy is wholly determined by one dimensional parameter, the **rate of kinetic energy decay** down the cascade of eddies, ϵ. The value of ϵ can be readily determined since $\epsilon = de/dt$, where e is the energy of oscillatory motion. The theory of turbulence 'omits' the mass m from the classic expression of energy, ML^2T^{-2}, so that $[\epsilon] = L^2T^{-3}$. Thus, in the range Λ to λ_m, the so-called **inertial subrange**, Λ to λ^*, can be singled out, in which energy is not lost during fractionation. The next range, λ^* to λ_m, covers eddies of smaller scale, where molecular friction can no longer be disregarded and a second parameter has to be taken into account – viscosity ν, already familiar to us. The corresponding interval is known as the **viscosity subrange**. The eddies of this subrange lose energy through dissipation to heat.

Repeating Kolmogorov's derivation of the expression for λ^*, the breakpoint scale separating the two ranges is easy to define. For this purpose, we will again make use of dimensional analysis. If λ^* depends on ϵ and ν, the dimension of each can be written as

$$[\epsilon] = L^2T^{-3} \qquad [\nu] = L^2T^{-1}$$

Now let

$$\lambda^* = \epsilon^\alpha \nu^\beta$$

where α and β are unknown. Substituting the dimensions into both sides of the equality, we obtain the relationship

$$L = L^{2\alpha} T^{-3\alpha} L^{2\beta} T^{-\beta}$$

Equating the exponents on the left and right, we find that $\alpha = -\frac{1}{4}$, $\beta = \frac{3}{4}$. Finally, Kolmogorov's scale is found from

$$\lambda^* = \left(\frac{\nu^3}{\epsilon}\right)^{\frac{1}{4}} \tag{4.2}$$

Knowing the viscosity ν of sea water and ϵ, we can calculate the scale λ^*, or the breakpoint between inertial and viscosity subranges. The calculation yields $\lambda^* \approx 1$ cm, showing that forces of molecular viscosity are only felt in an extremely narrow range of eddies.

Finally, another important consequence of the Kolmogorov approach should be discussed. Since in the range of scales from Λ to λ^* no energy of oscillatory motion is lost in the cascaded disintegration of eddies, it is possible to estimate the total energy of a set of eddies of the same scale. For the relationship to be derived, we need only to take into account one parameter, ϵ. Instead of energy $[e] = L^2T^{-2}$, turbulence theory deals with the variable S, $[S] = L^3T^{-2}$, which is called the energy spectrum. The scale l is replaced by its reciprocal, the wavenumber $k = 1/l$.

Again, making use of dimensions, we write

$$S(k) = c_1 \epsilon^\alpha k^\beta$$

or, in dimensional terms,

$$L^3T^{-2} = (L^2T^{-3})^\alpha (L^{-1})^\beta$$

Then we obtain

$$\alpha = 2/3 \qquad \beta = -5/3$$

This gives the well known Kolmogorov 'minus 5/3 law'

$$S(k) = c_1 \epsilon^{2/3} k^{-5/3} \tag{4.3}$$

where c_1 is a universal constant, the empiric value of which is 1.4.

Since turbulent eddies transfer any substance or momentum down a cascade of diminishing scales, the expression for the eddy coefficient for momentum, $A(l)$, can be derived by using the preceding procedure:

$$A(l) = c_2 \epsilon^{1/3} l^{4/3} \tag{4.4}$$

where $c_2 = 0.1$, as established by a great number of

experiments. This is another well known law of geophysical flows ('4/3 law'). Like local isotropy, it was discovered 'on paper' by A. Obukhov (1941), who arrived at it by means of complex calculations, as well as being found empirically by L. F. Richardson (1926), who observed the movement of smoke in the air during fairly ingenious experiments.

Equations 4.3 and 4.4 are helpful in clarifying the nature of turbulent processes. Suppose, first of all, that we have computed the energy spectrum $S(k)$ for a certain range of wavenumbers k. On a logarithmic plot of this function (Fig. 4.3), a range of wavenumbers may obey the 'minus 5/3 law'. This would indicate the presence of a cascaded energy transfer (inertial subrange). On the other hand, peaks and valleys on the curve would indicate the influx of outside energy or, conversely, energy efflux from the oscillatory field. Kolmogorov's theory has thus provided geophysicists with an important tool of research for identifying sources and sinks of energy of turbulent motion. A more rigorous description of Kolmogorov's ideas can be found in the book of Tennekes and Lumley

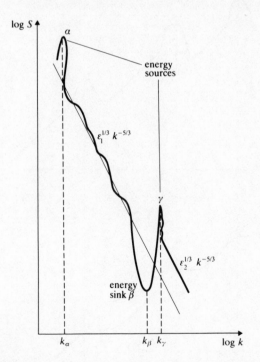

Figure 4.3 Hypothetical distribution of energy sources and sinks in the domain of wavenumbers k. The source γ (different from source α) is powerful enough to create a cascaded energy transfer with a higher energy dissipation rate ϵ_2 (ϵ_1).

(1972). The emergence of this theory was immediately followed by important discoveries, one of which was made by Ozmidov (1965a,b) in regard to oceanic turbulence and is discussed in the following section.

4.4 Sources of energy supplied to oceanic motions of different scales

As we know, the main current-generating forces are anisotropic in their nature. Indeed, the wind stress at the surface is horizontal. Owing to unequal heating of the ocean, a density gradient also occurs in the horizontal plane. Therefore, the large-scale structures in the current field induced by these external forces are basically anisotropic.

These gyres and eddies are unstable. As we know, the instability is measured by the Reynolds number, which is always large for such large-scale motions. This triggers the breakdown of eddies analogous to the Kolmogorov process. Up to what stage, however, will these 'filial' eddies (Ozmidov's term) retain their original horizontal orientation?

There should exist a critical point following which an eddy or turbulent fluctuation will 'forget' its original horizontal orientation. This scale of 'overturning' of the eddies, l_{cr}, was calculated by Ozmidov also by means of universal dimensional analysis. Obviously, the 'overturning' of the eddies will lag in the case of pronounced stratification $\partial\rho/\partial z$, and, conversely, will be facilitated by more intense cascaded energy transfer. This is the rationale behind Ozmidov's expression

$$l_{cr} = c_3 \left(\frac{\rho \epsilon^{2/3}}{g \dfrac{\partial \rho}{\partial z}} \right) \tag{4.5}$$

where $c_3 \sim 1.0$.

It seems appropriate here to give some estimates of the scale of 'overturning' of eddies. The value of ϵ can be assumed equal to 10^{-6} m^2 s^{-3}. By way of example, consider two density stratifications, $\partial\rho/\partial z = 10^{-4}$ kg m^{-4} (weak stratification) and $\partial\rho/\partial z = 10^{-1}$ kg m^{-4} (sharp stratification), and let $\rho \approx 1020$ kg m^3. Then in the former case, according to Equation 4.5, $l_{cr} = 32.6$ m, whereas in the latter case $l_{cr} = 0.18$ m. This gives an idea of the extent to which vertical stratification inhibits vertical turbulence. In uniform water a large (over 30 m) three-dimensional eddy is possible, whereas in sharply stratified water the same effect can only be achieved by 1 m or smaller whirls.

After this elegant proof of the predominantly horizontal orientation of the majority of turbulent motions in the ocean, Ozmidov proceeded to an analysis of the sources of energy of a turbulent field. The question was, what scales of oscillations can be induced in the ocean by external forces and, conversely, which oscillations are produced as a result of the Kolmogorov cascaded energy transfer? Following Kolmogorov, the energy spectra of currents for eddies of different scales were to be studied. Ozmidov started from a hypothetical scheme (Fig. 4.4) with a logarithmic plot of the energy spectrum S and the wavenumber k. For convenience, the appropriate scales are also shown in the figure.

Three ranges where the 5/3 law is observed are clearly shown by straight lines. At these scales the energy transfer down the cascade of splitting eddies entails no energy loss. These are segments of local isotropy, which refer to the horizontal plane, however, as there is no predominant direction in the evolution of the eddy in relation to the cardinal points.

On the curve in Figure 4.4a, peaks of energy influx are also seen. The first of these (peak a_1), which corresponds to a scale in the order of 10^6 m (1000 km), reflects the influx of energy from quasi-stationary cyclones and anticyclones (Fig. 1.5), which, as we know, produce the greatest inhomogeneities in the current and temperature fields in the ocean. Disturbances around the point a_1 create large-scale oceanic patterns – those powerful gyres described in Section 2.2.

Another feature of the curve $S(k)$, near the point a_2, corresponds to the scale 10^4 m (10 km). A disturbance of this kind in the current field may be due to the Coriolis force. A case in point is the inertial current (Fig. 3.7) arising in a water body

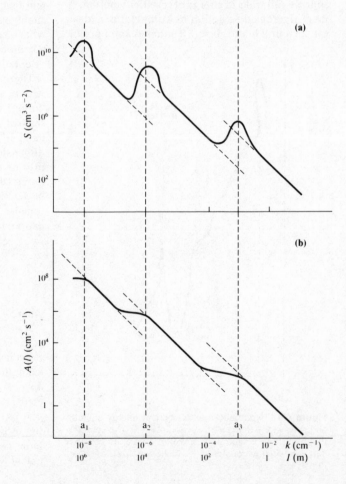

Figure 4.4 (a) Discrete energy supply of an oscillatory ocean field. (After Ozmidov 1965a.) (b) Distribution of eddy viscosity coefficient corresponding to diagram (a). (After Okubo and Ozmidov 1970.)

affected by no other forces. Tidal forces due to the Moon's and Sun's attractions can also create such regular fluctuations in the current field.

Finally, the third peak a_3 is clearly associated with the size of the wind-driven waves ($\sim$ 10 m). According to Equation 4.5, a segment of scale smaller than a_3 reveals isotropic three-dimensional motion. This is the area of small-scale ('time-scale') turbulence. The splitting process here evolves according to the pattern described for the scale range λ to λ^* in the Kolmogorov model. Eddies in the range between a_2 and a_3 are referred to as mesoscale turbulence.

Ozmidov attempted to verify his concept by calculations of actual spectral energy in the ocean. This proved to be a difficult undertaking. The theory of random processes implies continuous recording of currents over a long period of time. However, usually currents are measured with a fixed sample interval. A series of observations had to be manipulated in order to make the theory applicable to this purpose, which was successfully done by Ozmidov and his colleagues.

They found that the small-scale turbulence in the ocean is indeed adequately described by the 5/3 law. Numerous experiments staged at sea by Ozmidov, and later by others, indicated that the mesoscale turbulence zone also includes a segment of local isotropy. However, certain features on the segment a_1 to a_2 indicate the absence of cascaded energy transfer from the source of largest scale. On the one hand, eddies in this range of scales are hard to observe, because the required length of observation should be about several years, which is difficult to afford. On the other hand, this range of scales is contaminated by wave processes, particularly the global Rossby waves (which will be discussed later). They are of a totally different nature and their energy distribution follows a pattern differing from the 5/3 law.

Generally, Ozmidov's scheme is best fulfilled in the open ocean, away from the coast, where all incoming external energy comes from above. Closer to the shoreline, the number of external sources may be greater. Indeed, the average current begins to be affected by the topography of the bottom and the coast, and new sources of instability may arise. This writer (Tolmazin 1972), for instance, accidentally discovered such additional sources of energy of mesoscale turbulence when studying nearshore currents in the Black Sea – a further confirmation of the complex nature of turbulent processes in the ocean.

Another method of verifying Ozmidov's hypothesis would be to determine the coefficient of turbulent viscosity for various scales of l (or k) and test the 4/3 law for the respective local isotropy ranges. This was, in fact, carried out by Okubo and Ozmidov (1970). Both scientists put together numerous calculations of $A(l)$ based on data on current oscillations and diffusion of dye and drifters, plotting the data as shown in Figure 4.4b. It was found that, in zones of the scale of intense external energy influx, $A(l)$ moves from one 4/3 curve to another, corresponding to a different value of ϵ. Thus, Ozmidov's hypothesis received an independent experimental corroboration.

4.5 Vorticity or curl

Before introducing another new concept of fluid mechanics, let us once again consider a horizontal shear current. This time, however, we will refer to a real, familiar process. Imagine a small boat quietly floating on a slowly moving river (Fig. 4.5).

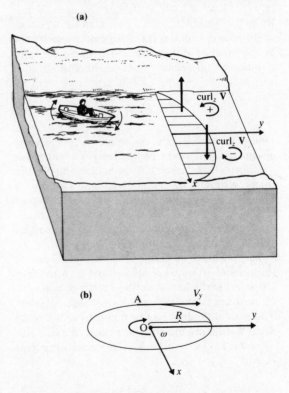

Figure 4.5 (a) Rotation of a boat in a shear flow (river current). Velocity profile and corresponding curl vector signs are shown on the right. (b) Net rotation of the bow, with no forward motion.

Obviously, the horizontal profile of the river flow will have its peak somewhere near the middle of the river since water particles are slowed down by friction against the banks. It can be readily seen that, in such a stream, the boat must start turning around, since from one side the water would flow into it at a higher velocity than from the other. The rotation will be the more intensive, the stronger the velocity gradient between the two ends of the boat. If, instead of a boat, we consider the volume of fluid expelled by it, that volume, as well as rotating, would sustain a change of shape. The tendency for fluid particles to change shape due to fluid rotation is called **vorticity** or **curl**.

Actually, the same effect is responsible for turbulence. However, at this point we are not concerned with the random eddies creating the turbulent fields, but we will now focus our attention on the behavior of the vector of the average current field.

Quantitatively, vorticity can be estimated by the gradient (shear) of velocity. The greater the shear, the stronger the tendency for rotation. The Greek letter ζ, or word 'curl' are frequently used (sometimes the symbol 'rot' is also used, an abbreviation of the Latin *rotatio*) to designate vorticity in geophysical fluid mechanics. Therefore, vorticity of a horizontal velocity field can be written as

$$\zeta = \text{curl}_z \mathbf{V} = \frac{\partial V_y}{\partial x} - \frac{\partial V_x}{\partial y} \qquad (4.6)$$

where V_x and V_y are the velocity components along the x and y axes, respectively. The subscript z indicates that the vorticity vector is normal to the plane of rotation. By convention, vorticity is assumed to be positive (the vector $\text{curl}_z \mathbf{V}$ goes along positive values of z) when rotation is counterclockwise (cyclonic) and negative for the opposite (anticyclonic) rotation (Fig. 4.5).

The physical meaning of vorticity is easy to understand if we look at the boat as a rotating particle. Let the bow describe a circle of radius R, eliminating translational downstream motion. The path of the bow can be depicted as in Figure 4.5b. We denote the angular velocity by ω. According to Equation 4.6, the velocity gradient at A is

$$\frac{\partial V_y}{\partial x} = \frac{2\omega R}{R} \qquad (4.7)$$

(the linear velocity $V = 2\omega R$ grows from 0 in the center as a function of the radius), since $V_x = 0$ on the axis Ox. Therefore, at any point of the circle, since the direction of the x and y axes was chosen arbitrarily,

$$\text{curl}_z \mathbf{V} = 2\omega$$

i.e. the vorticity of a solid body is equal to twice the angular velocity. Translating this result to the Earth's rotation (see Section 3.3 and Fig. 3.6, in particular), its curl is exactly equal to the parameter $f = 2\omega \sin \phi$, which is called **planetary vorticity**. For consistency, designation ζ_p will be used also, but the reader should remember that $\zeta_p = f$.

As we have seen on several occasions, rotation is a typical feature of currents, and curl, therefore, is a convenient expression for their spatial inhomogeneity. In striking against banks or bottom features, a water flow changes its shape and size, giving rise to longitudinal and transverse gradients that constitute $\text{curl}_z \mathbf{V}$.

Besides, fluid particles moving along the meridian, e.g. toward the Pole, migrate from areas of slow planetary rotation (in the horizontal plane!) into those where the speed of rotation is higher. (The variation of the Coriolis parameter with latitude and the model of inertial rotating current should be recalled here.) This also leads to a changing curl value.

Tops and gyroscopes are known to possess an amazing stability of motion and the moment of rotation proves to be no less conservative than energy or mass. For any change of rotation to occur, some external action is required. This also holds for fluid particles whose moment of rotation is measured by curl. For every liquid, it is possible to formulate its law of conservation of vorticity and the balance of influences increasing or decreasing the curl of its particles. This law is derived from the previously described equations of motion expressing the conservation of momentum.

4.6 Integral circulation and stream function

Since water is extremely mobile and fluid, it may seem that each of its particles can move unimpeded in any direction. At least, such is the impression of a swimmer, who meets equal resistance from all directions. The complete description of any such motion therefore requires

78

studying the forces in three-dimensional space. Thus far, our balances of forces and velocities have been computed only in the horizontal plane. The real picture, of course, is much more complicated. A closed system of fluid dynamics equations comprises three equations for the conservation of momentum (balance of forces) according to the three directions of the co-ordinate axes. Another equation serves to estimate pressure – an important internal force operating inside the fluid. In addition, there is an equation of state (linking density with temperature, salinity and pressure) and an equation describing the transport of heat and salts.

Even disregarding the processes forming the density structure of the ocean and assuming that this structure is known, we still have to deal with an extremely complex system of equations. It was only recently that efficient methods have been developed to solve this system using computers. The usual previous approach was to simplify the equations in the search for a clear and easy, although approximate, solution, which, in many cases, was enough to account for the large-scale oceanic circulation and some of its major characteristics.

We will now try to derive a number of simple specific conclusions from the complex general laws.

Before we do that, one important reminder: as we know, the ocean is no more than a thin film on the planet's surface. Its maximum depth does not exceed 1/600 of the Earth's radius. Besides, as we have seen in Section 4.2, the vertical stratification of the ocean impedes turbulent energy transfer downward, so that the principal large-scale motions are all concentrated in the layer above the thermocline, restricting the vertical dimension of the ocean surface currents even more.

Let us make a rough comparison of the scales of vertical and horizontal movements of fluid particles. This is readily done for tidal motions, whose vertical displacements are the fastest.

The highest tide on the Earth attains 18 m (Bay of Fundy, Nova Scotia). Fluid particles are brought to that height in 6 h with a vertical velocity of about 8×10^{-2} cm s^{-1}. Even with a lower tidal height, horizontal tidal currents attain speeds of 200–250 cm s^{-1} in narrows. A similar computation for the velocities of vertical water movements accompanying wind-driven currents in the open ocean shows that their average values do not exceed 10^{-5}–10^{-6} cm s^{-1}.

We thus see that the principal direction of movement of water in the ocean is horizontal. It is therefore possible, without introducing major error, to exclude from the analysis the vertical velocity and all the forces acting in the vertical direction except for gravity and the pressure gradient. When the latter two forces are balanced, the ocean is in a state of hydrostatic equilibrium: at any point the pressure is equal to the weight of the overlying unit water column – a law that we have extensively used earlier in this book (Section 3.2). Having thus discarded vertical motions – although we shall return to them later – we will now have to make do with just two velocity components. Equations determining their behavior in space, however, remain 'three-dimensional', allowing variation of all motion characteristics in three directions. Solution of these equations is still extremely complex.

In 1946, when a numeric solution of problems of this complexity was not yet possible since high-speed computers did not exist, the Soviet oceanographer, V. B. Shtockman (1946), proposed to exclude from analysis also details of the vertical structure by integrating velocities over the vertical through the entire water column.

Imagine a narrow vertical parallelepiped with a square base extending from the sea surface to the bottom and with lateral facets parallel to the co-ordinate axes. In order to evaluate the amount of water crossing these facets per unit time, we must add up the minute volumes of water flowing through all the infinitesimal areas making up the facets from the surface to the bottom. Each such volume is equal to the velocity multiplied by the area of the respective portion of the facet, and summation in this case is identical to integration. In symbolic notation, this appears as follows:

$$S_x = \int_0^H V_x \, dz \qquad S_y = \int_0^H V_y \, dz \qquad (4.8)$$

where S_x and S_y are components (projections) of the integral volume transport $\mathbf{S}$, V_x and V_y are projections of the current velocity, dz is an infinitesimal increment (element) of the vertical co-ordinate and H is the ocean depth.

If we assume that the parallelepiped has facets of unit width, the integral (total) volume transport is the specific (per unit length) flow rate.

Note that V_x and V_y may vary with depth in a complex pattern and may even reverse their sign.

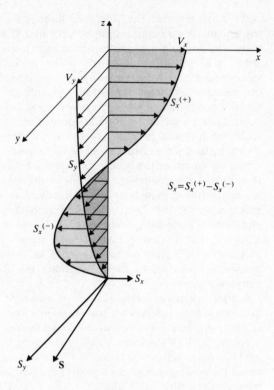

Figure 4.6 The procedure used to calculate the integral transport **S** by measured current vectors at different depths. Lines connect endpoints of vector projections on Ox and Oy axes. Estimation of S_x and S_y is reduced to calculation of the area bounded by these lines and the respective axes. Area on the plane xOz is hatched; $S_x^{(+)}$ and $S_x^{(-)}$ are indicated by light and dark shading. At the bottom is shown the intergral transport vector and projections in conventional units.

The corresponding products $V_x dz$ and $V_y dz$ are summed up algebraically (Fig. 4.6).

This expression laid the foundation for the **theory of integral circulation**, or 'total transport', which even today remains a major method of dynamic oceanography and has been used to account for important features of oceanic motions.

Integration eliminates the vertical co-ordinates from the equation of motion. The resulting relationships describe the motion of a fluid in a 'two-dimensional' space or a horizontal plane. Although this is, of course, not motion of individual particles but rather total transports embracing the entire water body, the system still remains fairly complex, as it comprises two equations for two components of the integral transport S_x and S_y.

We have seen earlier that two velocity components can be reduced to one vertical curl vorticity component, that is, two unknowns can be reduced to one. This operation is also performed for total transport equations, by applying the gradient operation† alternately to both, and subtracting one from the other according to Equation 4.6. As a result we can deal with a scalar quantity – the vertical component of the curl – and it is possible to write a balance equation for this quantity if the sources of gyrations are known.

One should not be overly amazed at this artifact. Like any other mathematical abstraction, it reflects certain properties of the real world. Specifically, it signifies that a fluid that rotates (and everything does rotate on this planet) tends to move uniformly in all the planes perpendicular to the axis of rotation, that is, in columns parallel to that axis. This tendency will, of course, take effect only when it is not impeded, and there are plenty of hindrances to this in a real ocean. Nevertheless, awareness of this tendency was crucial in the successful development of the theory of integral circulation.

It is not difficult to assemble the vorticity balance equation. As for any other quantity, we should define all 'sources' of gyration and all 'sinks' that reduce the tendency of water columns to rotate. Of course, we are interested in gyration relative to the Earth, and therefore the so-called **relative vorticity** or **relative curl** ζ_R is to be found. For the deep ocean, four principal components of the relative curl are well defined.

First, the irregular wind blowing over the ocean surface transmits some of its vorticity ζ_τ via tangential stress. It is obvious that the bottom stress should also affect the gyration of the water column. However, in the deep baroclinic ocean, the bottom stress can be neglected.

Secondly, if a current has a meridional component, the rotation of a water column is affected differently at various latitudes (Fig. 3.6). This effect was termed planetary vorticity ζ_p ($= f$), and will be described in more detail in the next chapter.

Thirdly, vorticity may increase or decrease when water columns move over irregular ocean beds. This is the so-called **topographic vorticity** ζ_T, which will be discussed in detail in Section 4.8.

Finally, part of the vorticity is scattered by horizontal turbulence at the lateral boundaries (shores or boundaries between oppositely moving cur-

† By gradient operation, or curl operation, we mean application of the rules for calculation of gradient or, respectively, curl to the variables or relationships concerned.

rents). These (horizontal) **lateral curl stresses** ζ_L largely depend on the scale of motion.

In short, the vorticity balance can be written as

$$\zeta_R = \zeta_\tau + \zeta_p + \zeta_T + \zeta_L \qquad (4.9)$$

The problem of oceanic circulation has thus been reduced to one equation describing variation of the vorticity or curl of integral transport in time and in the horizontal plane. It is a partial derivative equation, that is, it relates the relative variations of vorticity at different points in time and space, rather than their absolute magnitudes.

The methods of solving such equations are not simple. The behavior of the curl at the ocean boundaries is also unknown, and hence it is difficult to formulate the boundary conditions. The equation only helps to find the form or variation of the unknown quantity, while the boundary conditions allow one to define its absolute values as well. Apart from difficulties in formulating boundary conditions for curl, there are two additional inconveniences. First, the curl is a function of integral transport S_x and S_y, which appear explicitly in those terms describing the changes of the curl of integral transport due to translation of the water column in the horizontal direction. Equation 4.8 thus includes two unknown quantities instead of one. Secondly, the total transport field, which is most important for us, cannot be reconstructed from the vorticity field alone.

The difficulties can be surmounted if the total transport (its two components) is expressed in terms of one quantity, in the same way as, for instance, geostrophic current is expressed through pressure gradient. It is possible, in fact, to derive a quantity ψ such that the total transport at any point would be equal to the gradient of ψ in the transverse direction, i.e.

$$S_x = -\,\partial\psi/\partial y$$

$$(4.10)$$

$$S_y = +\,\partial\psi/\partial x$$

Following exactly the same reasoning as we used for the geostrophic motion, one can see that the vector of total transport **S** is directed everywhere along the lines of equal values of ψ and that the larger values of ψ are on the right side.

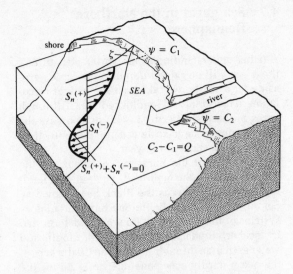

Figure 4.7 Boundary conditions on the shoreline. Total transport along the normal to the shore, S_n, is equal to zero: the drift component $S_n^{(+)}$ is balanced by the gradient transfer $S_n^{(-)}$ as a result of the surface slope $\partial\zeta/\partial n$ (ζ being the surface elevation of the undisturbed level). The shoreline coincides with the stream function, which abruptly changes its value at the points where the river enters the sea (Q is river discharge).

Now in the vorticity equation the scalar function ψ can be used instead of S_x and S_y, and it becomes an equation for a single **integral stream function** ψ.

The advantage of this equation, which is still rather complex, is that it allows one to find more easily boundary conditions for ψ that are physically correct. Since oceans are bounded by coasts, and water can neither accumulate near the coast nor cross it[†] (Fig. 4.7), the streamline near the shore will be parallel to it. Since the isolines of ψ are streamlines, the value of ψ should not vary along the shore. In other words, the shoreline is a streamline. The field of the integral stream function ψ thus provides a very general description of oceanic water motion. The isolines of this field are the streamlines along which vertical columns of water move. The equation that relates the function ψ to the curl of the wind stress and the variable depth and density structure of the water allows computation of the fields of total transports in the World Ocean. However, to begin with, a few simple examples will be helpful.

---

† We disregard, of course, smaller-scale phenomena such as occasional flooding of the sea shore.

4.7 Sea gyres in the Northern Hemisphere

A better understanding of the vortex structure of the ocean will be gained if, putting oceanic gyres aside for a while, we take a closer look at circulation in marginal and landlocked seas. Although gyres in these smaller basins do not possess the permanency of large-scale oceanic currents, they do persist for most of the year.

The basins of the Eurasian continent are the deepest cut into the mainland. Such almost landlocked water bodies as the Black Sea, Azov Sea, Sea of Marmara and Baltic Sea have circulatory systems of their own that are wholly determined by local driving factors – winds, river runoffs and water exchange through straits. The tidal waves of the ocean hardly ever penetrate here. As for the Caspian Sea, it is basically a large salt lake. All these basins have one remarkable common feature: the average circulation of water in them is typically cyclonic (Fig. 4.8). For about three-fourths of

the year, the water circulates in the counterclockwise direction. For the rest of the year, the circulation is either so weak it has no identifiable predominant direction, or appears as a weak anticyclonic circulation persisting for a short period of time. The schemes of currents in such marginal seas as the White Sea, Sea of Okhotsk, Sea of Japan, Yellow Sea, North Sea and others, and mediterranean seas such as the Red Sea, European Mediterranean Sea (with all its inland seas) and the East China Sea also show fairly stable cyclonic systems. Tidal currents from the open ocean come into some of these seas, but this does not affect their internal wind-driven circulation to any significant degree. In other seas, branches of ocean currents make up part of their gradient current systems.

Details of current systems in individual seas are beyond the scope of this book, which is concerned with the general dynamics of currents. These details are studied in texts of regional oceanography (see, for example, Dietrich & Kalle 1963).

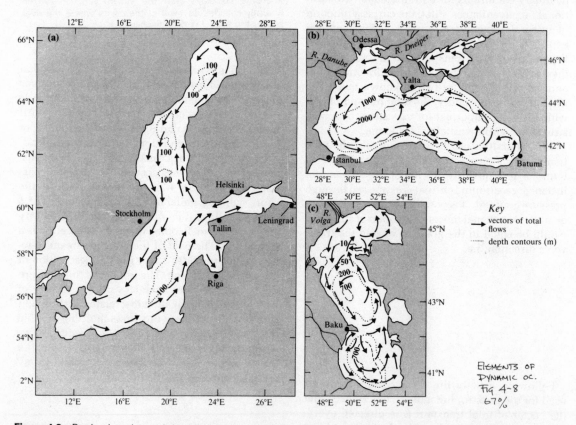

Figure 4.8 Predominant integral circulation in landlocked seas: (a) the Baltic Sea; (b) the Black and Azov Seas; and (c) the Caspian Sea. Depth is given in meters. (Generalized from Shamraev and Shishkina 1980 and Leonov 1966.)

Here we will try to explain one basic fact – the predominantly cyclonic gyration in the seas of the Northern Hemisphere.

Consider an abstract model of a sea so deep that potential energy is accumulated in it as a result of surface slopes and horizontal density gradients. It can be assumed that, because of the small spatial scale of the sea, the variations of the Coriolis parameter are so small that the planetary curl is negligible compared to other vorticities. As a result, we obtain Shtockman's model, described above.

We assume that this sea has an arbitrary shape and that wind blows over it in one direction with a transverse velocity shear (Fig. 4.9). For convenience, wind systems are shown in the figure as a field of isobars at a certain altitude where their regime is close to geostrophic. At the ground, the wind vectors will be deflected from isobars due to friction and their magnitude will be slightly attenuated. Here this is irrelevant, however, for we can reasonably assume that the transverse wind shear will remain unchanged, so the vorticities of the geostrophic wind and of the ground wind will be roughly equal. Disregarding for now also the variation in height of the water columns due to bottom topography, let us consider, in qualitative terms, what total transport field will be generated by a wind associated with the pressure field p_i ($i = 1, 2, \ldots$).

In the area of intense wind (in the eastern part of our test basin), the drift currents will obviously penetrate to a greater depth than in other areas. This drift transport will pile up surface water partially to the right, partially in the direction of the wind. This will create barotropic currents that will turn total flow to the left. The density field will also adjust to currents. The deep dense water will upwell against the western coast, causing the barotropic gradient G_2 (Fig. 3.8) which induces deep geostrophic flow opposite to the wind-induced transport. This means that, in the eastern part of the sea, integral transport will occur in the direction of the wind. The continuity of volume requires that this transport be balanced by influx of water from other regions. As a result, in the west the integral transport will be directed against the predominant wind. This means that under the isobars p_1, p_2, p_3 drift currents are negligibly small but the total transport is caused by the surface slope and density gradients along the normals to the coast. The resulting scheme of the stream functions ψ_i is shown in Figure 4.9. One can easily visualize that the picture of integral circulation will remain unchanged if the system of isobars is rotated through 90° with the same velocity magnitudes. Shtockman has, in fact, proved this quite rigorously.

The balance of vorticities of average vertical transport is as follows. Horizontally, non-uniform wind creates vorticity by the force of wind stress on the water surface. The loss of vorticity (when bottom friction is unimportant) should occur at the lateral boundaries of the sea. The contribution of lateral (horizontal) dissipation of vorticity by turbulent eddies in this case is particularly strong.

The next step taken by Shtockman was to show that integral circulation in the imaginary sea of Figure 4.9 would not be altered if, instead of a

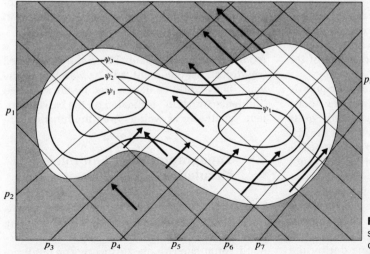

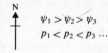

Figure 4.9 Two isobar systems, geostrophic wind with horizontal shear and corresponding total transport in a sea of arbitrary shape.

rectilinear pressure field of the type p_i, it were subjected to a closed rotating system of winds. One limitation, however, is indispensable: the horizontal shear of wind velocity and its average magnitude should remain unchanged. These characteristics of Shtockman's model provide an adequate explanation for the predominant cyclonic circulations in the seas of the Northern Hemisphere.

Any viewer of televised weather reports is bound to wonder how all those ever-changing low- and high-pressure systems could produce anything like steady, sometimes stationary, motion in the sea. The centers of baric systems pass alternately through the middle and along the border of the sea, just touching it. How can the energies of these extremely diverse wind patterns be accumulated in the sea to produce one or two simple gyres? This is, in fact, what happens – the stationary regime arises as the cumulative effects of a great number of events. In this sense, average conditions produce a result equivalent to that of permanent wind forces. As for the fast changes of lows and highs, the frequency of occurrence of baric patterns is less important here than the intensity of the characteristic shear (vorticity) of the wind that they generate.

An enclosed landlocked sea – Black, Baltic, Azov and Caspian Seas (Fig. 4.8) – is smaller than the majority of atmospheric lows and highs and, as often as not, it happens to lie at the edge of an atmospheric pattern rather than near its center. Instead of a circular wind system, skewed wind flows with a transverse cyclonic or anticyclonic shear are thus created over the sea, each generating a more or less identical integral circulation of water, regardless of which side of the cyclone or anticyclone actually hits it. According to Shtockman, their effect on the currents will be virtually the same. It is important, however, which of the two patterns is predominant.

Low-pressure systems pass quite frequently over Europe. The wind always has a high velocity and the transverse gradients are generally quite strong. High-pressure cells are usually steady, retaining the same position for several days. The gradients of anticyclonic fields are usually not large since these cells themselves extend over enormous areas. The effects of the numerous low-pressure cells, superimposed on one another, result in the prevalence of cyclonic circulation in the seas (Fig. 4.8). In the Black Sea, for instance, it is observed for over 70% of the time during the

year. However, the opposite gyration is also observed in winter and summer, the seasons of stable anticyclones.

Let us now return to the abstract models illustrated by Figures 4.6 and 4.9, which helped us to understand the origins of integral circulation of various vertical water structures. Recall that we began our reasoning from vertical integration of velocity components along x and y axes. By this operation we obtained the component of total transport. However, an inverse transition – from the transport stream functions (or the field ψ) to the description of currents in the vertical plane – is impossible: using Shtockman's model, we can only calculate the field ψ from a known wind field. In reality, integral circulation will not necessarily be ideally smooth as in theory. In fact, a gyre may be stretched or even broken.

Soon after the creation of the theory of integral transport, some modifications of it appeared, which attempted to describe the density field as well as integral circulation.

One fairly representative characteristic of the density field is the **baroclinicity depth** H. R. O. Reid (1948) was the first to investigate the density field jointly with integral circulation. He suggested a 'rigid' model (form) of the vertical density distribution, which was capable of horizontal evolution in response to varying boundary conditions and total transport. In Reid's model, the baroclinic layer to which the integration was carried out was not known and was determined in the course of the solution, as was the field ψ. Reid obtained a fairly good match between observed and calculated water transport values in the equatorial Pacific. This result aroused oceanographers' interest in density models for a certain period of time.

Later Shtockman (1952) himself developed a model of the density field. This field was defined 'rigidly' at just one point, and its variation in the water body could be determined by the so-called 'influence function', estimated on the basis of certain hypotheses as well as from the function ψ.

4.8 Topographic effects in integral circulation

The effects of the container's geometry on currents in pipes, open canals, flumes and rivers have long been known, certainly since 1768 when D. Bernoulli formulated his classic theorem on unidirectional flows. It stated that current velocity is inversely proportional to the cross-sectional area of the flow. Following this theorem, some more intricate flow characteristics such as streamlines and trajectories can be related to the bottom topography of a flow bed. If you observe a floating object in a brook, you may soon discover that its curvilinear path is closely related to small and large topographic features (say, stones and holes) on the brook's bed. The way in which bottom topography affects the complex wind-driven or density currents at sea has been studied only relatively recently, and topographic vorticity in the ocean and coastal waters is the subject of very active research.

So far, we have not discussed the effects of bottom topography on vorticity. In fact, the total transport in the deep ocean is only insignificantly affected by the ragged sea bed. However, in shallow places the role of bottom geometry may become dominant in shaping the flow patterns. Close proximity of bottom floor enhances the effect of turbulent friction through the increased vertical velocity shear. Another set of forces is associated with variations in the height of moving water

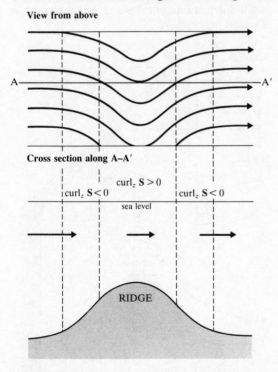

View from above

Cross section along A–A'

curl$_z$ **S** < 0   curl$_z$ **S** > 0   curl$_z$ **S** < 0

sea level

RIDGE

Figure 4.10 Behavior of the total transport stream functions over a long bottom ridge. (Adapted from Shtockman 1948a.)

columns: the average flow displays local accelerations or decelerations when a flow vertically shrinks or stretches over shallow or deep places. In such cases horizontal velocity gradients may produce horizontal friction. Flow fields, affected by the basin's geometry, are usually quite complex owing to interplay between the forces of friction, local accelerations and the Coriolis force. These effects are simultaneously at work in tide-driven estuarine currents. Wind-driven and gradient currents in the open ocean have generally simpler structures even near prominent topographic features. There were several attempts to isolate various circulatory mechanisms of topographic origin.

If a homogeneous water body is so deep that the friction along the vertical becomes insignificant, vorticity may exist owing to shrinking or stretching of water columns over irregular bottom floor. In this case, the sign of curl (rotation trend) is determined by the curvature of topographic features. Shtockman (1948a) proved that a current over a long ridge will display negative vorticity before and after the ridge (Fig. 4.10). Above the ridge's crest, vorticity will be positive, and the entire flow pattern will show U-shaped streamlines. The topographically related circulatory features in current fields are frequently referred to as the **topographic effects**.

Shtockman (1948a) showed that the interaction between the lateral shear due to non-uniform depth and the vorticity due to surface-to-bottom friction (deformation of the Ekman spiral at the transition from shallow to deep water) greatly depends on the total depth. Specifically, the latter effect may be discounted in a sufficiently deep sea. But how deep is deep enough? This is a much more complicated question than it sounds.

A. Felsenbaum (1960) proposed a convenient classification of seas by relative depth. A sea or other water basin in which substantial drift currents penetrate to the bottom is called 'shallow' or a **low-depth sea**. Its depth h is much smaller than the Ekman layer D ($h \ll D$). The velocity gradient here is so large that the Coriolis force is negligible compared to turbulent friction.

If the sea depth is less than $2D$ ($h < 2D$), the gradient current below the Ekman layer experiences strong effects of both friction and Coriolis force throughout the water body. This is a **medium-depth sea**, in Felsenbaum's classification.

Finally, in a **deep sea**, i.e. the kind of sea described above in connection with the Ekman model (Section 3.7), only the thin near-bottom layer of gradient current is affected by friction, while throughout the water body the friction forces are vanishingly small compared to the Coriolis force. As we know, the wind-driven current field in a homogeneous sea is different from that in the real ocean, but in a shallow or medium-depth sea, stratification may be quite insignificant. As a result, the barotropic model is suitable for the study of currents in a great number of seas and bays.

Elaborating on the Ekman method, Felsenbaum reduced the complex problem to a form in which it was conveniently solved by using the integral stream function ψ. Up to the determination of S_x and S_y all the operations were carried out according to the Ekman model: velocity was determined depending on the surface slope and then the integral transport was found. Felsenbaum's innovation was that he formulated a rigorous method obtaining a single equation for ψ with convenient boundary conditions. His model allowed him the inverse operation, impossible in the Shtockman model – to determine the velocity field from the computed stream function ψ. It became possible because all functions S_x, S_y, V_x, V_y and surface slopes were related by analytical expressions.

Let us now discuss specifics of motion in a shallow sea, beginning, as usual, with the structure of individual currents and only then with the total transport.

Operating with our familiar mechanisms – the pressure gradient and the wind stress – let us take the elementary case of a horizontally uniform wind blowing over an unenclosed sea, bounded only at its windward margin, and with a flat bottom. In this case neither topographic effects nor any of the driving factors will generate vorticity. Because the sea is very shallow, the deflecting force of the Earth's rotation is insignificant compared to friction, and the current in the upper layers will be directed along the wind. As a result of this transport, the water will pile up near the shore and a pressure gradient will set up in the water column that will be oriented directly opposite to the wind. The velocity vector at the surface produced by the wind will decrease rapidly with depth and will turn through 180°. Since the water particles are assumed to adhere to the sea bed, the velocity of the deep countercurrent will reach its maximum below the surface onshore current but above the sea bed. Figure 4.11 shows the velocity profile for just one point at sea, but it should be identical everywhere.

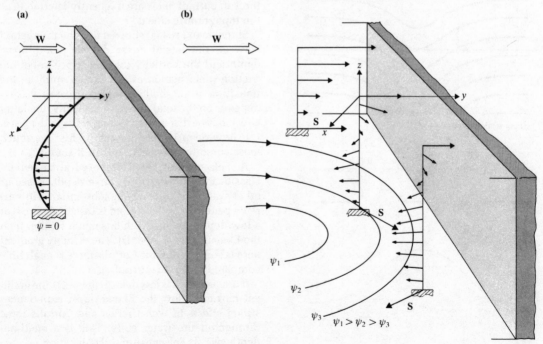

Figure 4.11 Circulation in a shallow sea: (a) a featureless bottom floor; (b) the bottom slopes in the direction transverse to the wind vector.

86

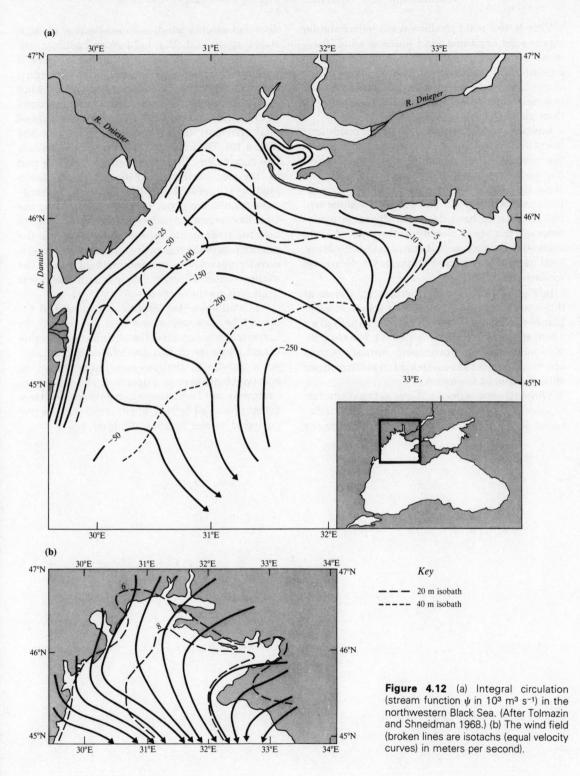

Figure 4.12 (a) Integral circulation (stream function ψ in 10^3 m³ s⁻¹) in the northwestern Black Sea. (After Tolmazin and Shneidman 1968.) (b) The wind field (broken lines are isotachs (equal velocity curves) in meters per second).

This is the main shallow-water effect: under steady-state conditions and uniform wind, there must be a countercurrent at a certain depth beneath each such current. The volumes of the upper and lower currents are equal and there is no integral water transport. In this case, $\psi = 0$ throughout the sea.

Imagine now that, with all other conditions being unchanged, the sea floor slopes transverse to the wind direction. In shallow areas (the farthest corner), the upper current then may be able to reach the bottom, while in deeper areas the countercurrent will prevail. The sea surface slope will not necessarily have the same direction everywhere. If the sea is closed (on the coast, $\psi = 0$), integral circulation will develop: in shallow areas, total transport is directed along, and in deep areas against, the wind.

In Figure 4.11b, velocity vectors are shown at three points. Estimating the total transport as in Figure 4.6, the resultant vectors S can be plotted on a conventional scale as in Figure 4.11. In a sea with a non-uniform bed topographic vorticity is thus observed, shown by the system of stream functions at the bottom of the figure.

With this simple model, it was easy to visualize a potential picture of the motion in qualitative terms. In a real situation with a non-uniform sea floor and varying wind, and the effects of springs, rivers, straits and other influxes and sources, circulation may be extremely complex, making it hard to isolate the contribution of each individual factor. The system of ψ lines for the average wind system clearly correlates with the inhomogeneity of the wind field. This is illustrated by a field of total transport in the northwestern Black Sea (Fig. 4.12). The influence of bottom topography is less significant. Three rivers empty into this part of the sea: Dnieper (1750 $m^3\,s^{-1}$), Dniester (400 $m^3\,s^{-1}$) and Danube (6300 $m^3\,s^{-1}$) – average river runoffs are given in parentheses. Only the Danube exercises any effect on the water dynamics near the river mouth. It is remarkable that in the eastern part of the region the total transport moves directly opposite to the predominant wind. The shallow-water effect in this example manifests itself only near the coast.

Can shallow-water conditions be found in the ocean? As we know, at and near the Equator the horizontal component of the Coriolis force is close to zero while the Ekman depth D tends to infinity. As a result, a 'shallow-water effect' may be observed there even in a deep sea. The subsurface Cromwell and Lomonosov Currents are to a large extent produced by this effect, which will be discussed at greater length in Chapter 6.

5 Westward intensification of surface currents

5.1 Mysteries of the Gulf Stream

All the great civilizations of the past arose in lands protected from the cold, the eternal foe of life. It is true that more recent European civilization originated on a continent less endowed with the warmth of the Sun, but it, too, seemed to be protected from the pernicious breath of the north by an invisible friendly hand. Since time immemorial, a mysterious force has twisted the air over the Atlantic Ocean into huge eddies – cyclones that have brought warmth and lifegiving moisture to the continent. On their way over Europe, cyclones have driven the biting frost and the searing heat deep into the interior of the Asian continent. For centuries, Europeans had no idea of the source of their mellow climate, and it was only after the beginning of the exploration of America that they met their benefactor face to face. It was the Gulf Stream – the powerful warm current that cuts across the North Atlantic from south-west to north-east.

Seamen invented a variety of names for this main Atlantic current, but only one survived – the Gulf Stream, the stream that starts at the Gulf of Mexico. In the 1870s this name became official, after Benjamin Franklin put it on his geographic map.

Diverse hypotheses as to the causes of the Gulf Stream have been proposed since its discovery. Some of its properties have been explained only recently, with the advent of the modern theory of oceanic currents, while others remain unexplained even today.

What was it about the Gulf Stream that baffled its explorers so much? First of all, they could not understand from whence came the enormous amount of water flowing from the Gulf of Mexico through the Strait of Florida. This mystery was clarified only when the general scheme of circulation of water in the Atlantic Ocean became known.

The protruding Brazilian 'corner' of South America forces not only the North Equatorial Current into the Caribbean Sea, but also a substantial part of the water of the South Equatorial Current, which passes through the Lesser Antilles where it is known as the Guiana Current (Fig. 2.2). Driven by trade winds, these currents merge, squeeze through the Yucatan Channel between Cuba and Mexico, flow west across the southern Gulf of Mexico, and thence, through the Strait of Florida, into the Atlantic Ocean. The Gulf Stream does not thus fully justify its name: the water of the Gulf of Mexico does not really contribute much to the fast-flowing stream crossing the gulf's southern end.

Seen in the general framework of the entire system of Atlantic currents, the origin of the Gulf Stream becomes much more lucid: the current forms part of the northern subtropical gyre and is also supported by waters from the Southern Hemisphere. However, what makes the Gulf Stream retain the form of a narrow jet in the ocean for hundreds of kilometers? From fluid dynamics – and from everyday experience – we known that a fast stream flowing out of a narrow container (be it a pipe, a channel, or a river) into a large basin expands and loses velocity rapidly. The stability of the Gulf Stream thus represents one of the paradoxes of circulation in the North Atlantic.

Another unanswered question was the persistence with which the Gulf Stream, on its way to the north, keeps clinging to the western offshore slope. In fact, one would expect the Coriolis force to deflect the stream to the right, that is to the south, immediately after it leaves the Strait of Florida. In reality, however, the stream takes the opposite course.

And, the final enigma – on passing Cape Hatteras, the Gulf Stream for no apparent reason makes an abrupt turn into the open ocean.

Many explorers have struggled to unravel these three paradoxes of large-scale circulation in the North Atlantic, but a satisfactory explanation was

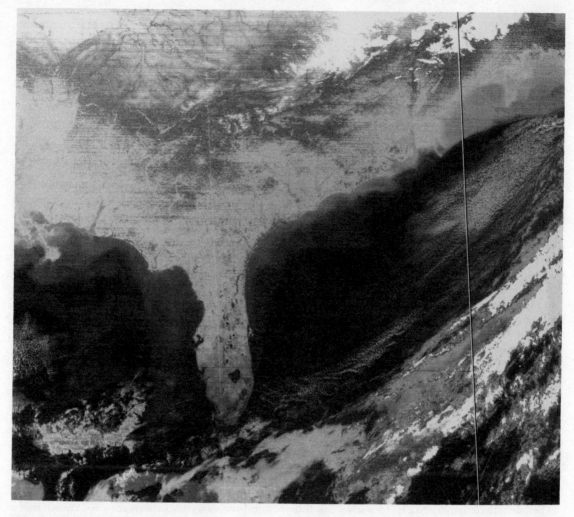

Figure 5.1 High-resolution infra-red VHRR image of the Gulf Stream (dark band) received on January 7, 1977 from NOAA-5 orbit 1287. (Courtesy of R. Legeckis, National Earth Satellite Service.)

only found in the late 1940s and early 1950s, when the theory of integral circulation appeared to be a convenient tool in dynamic oceanography, as discussed in the preceding chapter.

The mysteries of the Gulf Stream have in fact forced oceanographers to develop new strategies of field studies. The earlier prevailing view was that an oceanic process develops over a long enough period of time to be investigated from a single ship traveling from one distant point to another. This method proved unacceptable for the Gulf Stream, where, apart from the inconvenience of measuring the velocity of a fast-moving current from a moored ship, the field of physical characteristics changes with extreme rapidity. It was decided, therefore, to use the properties of the Gulf

Stream itself, taking advantage of its being saltier and warmer than surrounding waters.

From the 1930s on, American explorers of the Gulf Stream, using several ships, repeatedly crossed it at many different spots. Following the invention of the bathythermograph (Section 2.5), it became possible in just a few days to perform a detailed temperature survey of the Gulf Stream. However, waters in the region are so dynamic that even this proved to be too slow. In 1952, Americans first measured the long-wave (thermal) radiation of the sea surface from a low-flying aircraft, obtaining an almost instant overall picture of the current.

Direct measurements of current velocities had been conducted since 1949 using the geomagnetic

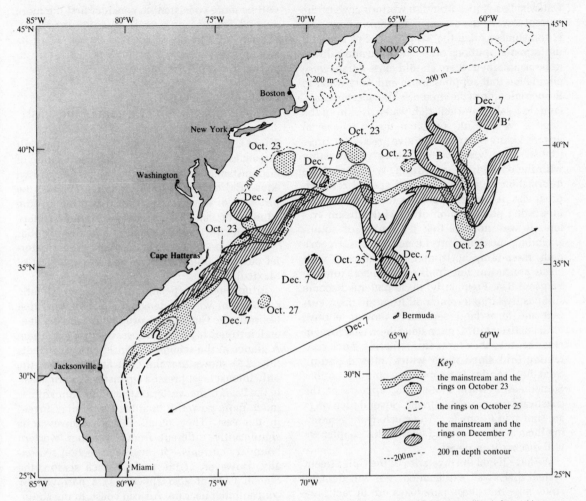

Figure 5.2 Positions of the Gulf Stream during two successive surveys obtained by NOAA-5 satellite infra-red imagery and bathythermographic data in October and December, 1978.

electrokinetograph (GEK), described in Section 2.5. In conjunction with deep oceanographic measurements (cross sections), this method yielded ample quantitative data on currents in the Gulf Stream region.

It was established that the Gulf Stream is not a uniform 'river in the ocean', but consists of several intermittent jets moving with different velocities. The maximum registered velocity is 2.7 m s^{-1}, which is, in fact, the highest current velocity observed in the open ocean.

Narrow countercurrents have been discovered on either side of the stream's core, with the seaward countercurrent being more stable and having velocities up to 1 m s^{-1}.

At present, the Gulf Stream system is continually monitored by various US agencies. The pos-

ition of the stream and of the eddies generated at its margins are detected by satellite infra-red imagery (Fig. 5.1), bathythermographic data, weekly current charts drawn using drifting sea buoys, and other measurements. The overall boundaries of the system and its individual flows and eddies are carefully mapped for the surveyed periods. Monthly charts of temperature fields and the stream's position covering the region between Florida and the 55°W meridian are published by the National Oceanic and Atmospheric Administration (NOAA) (Fig. 5.2). Weekly temperature charts are also published by the US Naval Oceanographic Office.

The stream separates the cooler coastal waters of North America from the warm and salty waters of the Sargasso Sea. The boundary between the two water bodies slopes from the western edge of the current toward the ocean. This sloping boundary is sometimes called the 'cold wall'. For the non-uniform distributions of temperature and salinity to be maintained, there should exist a transverse circulation that supplies warm and salty water to the oceanic side of the stream, while near the coast, cold, less salty, nutrient-rich water rises from the depths. As a result, plankton thrive in coastal waters, giving them a distinctive green-blue tinge (the so-called 'green water'), sharply contrasting with the dark-blue Gulf Stream. With the ample information now available and special research programs, it became possible to investigate an interesting phenomenon of the Gulf Stream system. It was noticed that a number of south-extending meanders are formed on the stream's path, resembling horizontal wave motions with crests protruding hundreds of kilometers into the Sargasso Sea. Frequently, these patterns become sufficiently large to pinch off from the main current and form large separately moving eddies. F. C. Fuglister (1972) has named them **Gulf Stream rings** because they form close rotating loops containing cold slope water which, after departure from the meander, is trapped in a circular motion. Sometimes a warm water ring appears at the coastward margin of the stream. Several such rings are shown in Figure 5.2. The mechanism generating them is important and later in this chapter we will discuss it in detail.

We have listed the mysteries of the Gulf Stream system that await explanation. We will demonstrate now that these paradoxes are in fact law-governed regularities. The reader, aware of various scales of oceanic phenomena, may question the consistency of this approach. Indeed, have we not decided from the start to disregard the small-scale movements and focus attention on events of a larger scale? The formation of meanders and rings certainly is a small process compared to the general dimensions of the oceanic gyres. Yet, if we compare the transverse dimension of the Gulf Stream with the diameter of the eddies, the difference would not appear all that significant. On top of that, we are not solely concerned with the behavior of currents as determined by certain specific factors, but rather are trying to trace the flow of energy from one system to another. The role of meanders and rings in scattering the energy of the ocean's currents is no less significant, however, than that of turbulence.

All this will be explained later, but for now it will be more consistent to consider first the more general features of overall oceanic circulation, subsequently proceeding to the finer traits of its dynamic structure. To begin with, let us try to answer the following question.

5.2 Why is the Gulf Stream in the west?

There is more than one reason to regard the Gulf Stream as the product of a peculiar combination of circumstances. These include the merger of two equatorial currents (northern and southern), the presence of a 'rigid' meridional boundary – the Isthmus of Panama (no such western boundary exists in the Pacific Ocean) – and the peculiar 'storage basin' of the waters carried by the equatorial currents (the Caribbean Sea and the Gulf of Mexico).

Admitting these peculiarities of North Atlantic circulation, we can demonstrate, however, that basically the Gulf Stream originates from the general, intrinsic regularities observed in every ocean. A glance at the scheme of World Ocean currents (Fig. 2.2) shows that the Gulf Stream is not the only intensive jet pressed against a western coast. In the Pacific, the western current, the Kuroshio, is much more powerful than the California Current in the east. The Somali Current, however, is significantly different from the other western boundary currents – it reverses direction seasonally. However, at the peak of each season, the Somali Current also appears as a narrow flow concentrated near the African coast. In the southern parts of the oceans, the difference between the currents at the western and eastern coasts is not so striking, but the Brazil Current is stronger than the Benguela Current. The warm Agulhas Current, known since Vasco da Gama's time, has a fairly steady flow while its counterpart in the eastern Indian Ocean is so unremarkable that it lacks a commonly accepted designation. The South Pacific Ocean is the only exception to this general pattern. The weaker East Australia Current, groping its way through the islands in the ocean's west, cannot be compared to the mighty Peru (or Humboldt) Current in the east.

There should thus be causes other than local or regional features of seabed or coastal configurations responsible for the compression of the Gulf Stream into a narrow, fast-moving strip at the western margin of the Atlantic. Indeed, the Gulf Stream would have existed even if there were no

West Indies, Isthmus of Panama, or Strait of Florida, because every current has a tendency for intensification at the western coast of an ocean, whatever its specific traits. This phenomenon has been termed 'the westward intensification of boundary currents'.

Before proceeding to its physical interpretation, let us see how this phenomenon corresponds to the models of sea gyres described earlier. We know that a circular wind or a wind with transverse shear induces sea currents that form a symmetric gyre around the 'center' of the basin. It will be recalled that, for instance, in an idealized sea (Figs 4.8 & 9) cyclonic winds create a closed integral circulation where the specific features of the sea – whether of uniform density or stratified – are irrelevant. How does this relate to the current field in a real ocean?

Figure 5.3 schematizes the wind streamlines over the North Atlantic. The lines are approximately symmetric with respect to the 'center' of the northern part of the ocean. We may expect that the current field generated by these winds would also be symmetric with respect to this 'center', as shown in Figure 5.3a. In reality, however, currents form gyres displaced to the west of the wind circulation (Fig. 5.3b), forming a 'gulf stream' at the western coast of each ocean.

The detailed explanation for the westward intensification of currents based on vorticity balance will be given in the next section. Now, as a first approach to this phenomenon, we discuss the effect of asymmetry of a subtropical gyre in the Northern Hemisphere using simple arguments of momentum balance.

The northern and southern sides of a subtropical gyre are driven differently by the trade winds and westerlies. The equatorial drift induced by the trade winds is located in the belts of low latitudes. Since the Coriolis effect is insignificant here (f is small), the entire current is predominantly directed along the wind, and tranverse Ekman transport (to the north) is small. On the northern side of the gyre, where the westerlies act upon the water, the Coriolis deflection is much greater, and one may expect a prevailing total drift transport directed to the south. Some surface water is accumulated between the two flows, and a mound is formed there with a crest that is displaced southward from the center of the gyre (Fig. 5.4). The induced surface slope creates pressure gradient G_1 which facilitates surface flows in the directions of the winds. As in the Lineikin model

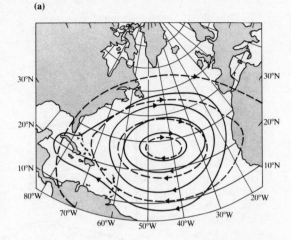

(a)

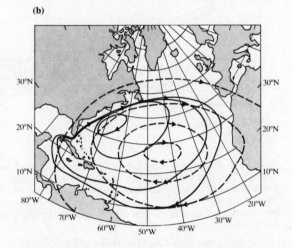

(b)

Figure 5.3 Schematic wind system (broken lines) and currents (solid lines) in the North Atlantic (a) for no asymmetry and (b) with asymmetry in the circulation.

(Fig. 3.15), the density field adjusts to this circulation, but now the isopycnals are not symmetric relative to the gyre's center, as is shown in the transverse plane (Fig. 5.4).

Thus, the system of westerlies and trade winds with north–south symmetry induces zonal flows with prevailing southward transport in the surface layer. This circulation would occur in a system without meridional boundaries (a zonal channel) non-uniformly rotating with the Earth. If a meridional boundary is present in such a channel (Sverdrup (1947) was the first to investigate this model), the northern flow is completely deflected southward. The mid-latitude ridge becomes asymmetric

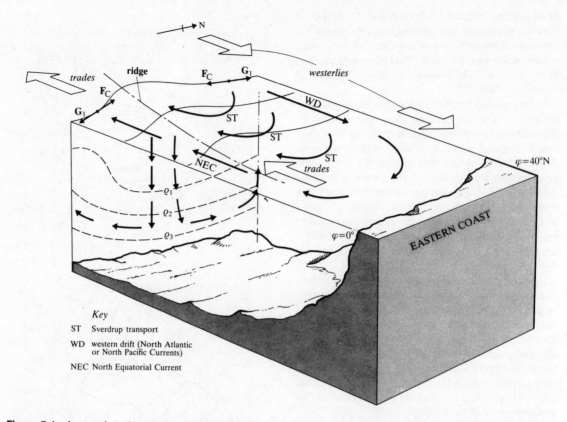

Figure 5.4 A part of a subtropical gyre in the Northern Hemisphere illustrating the mechanism of Sverdrup transport. *Key:* ST, Sverdrup transport; WD, western drift (North Atlantic or North Pacific Current); NEC, North Equatorial Current. The balance of barotropic pressure gradient G_1 and Coriolis force F_C is shown at each slope. The vertical motions (thin arrows) of isopycnals ρ_1, ρ_2 ... are shown on the back transverse plane (broken lines). Streamlines of Sverdrup transport are shown at the surface, but they actually represent the total transport from the surface to the depth where the distortion of isopycnals is negligible.

in the east–west direction: the ridge grows higher westward, as shown in Figure 5.4. An additional pressure gradient reinforces the southward transport. This so-called **Sverdrup transport** is a key element in ocean circulation.

The streamlines of Sverdrup transport, consisting of wind-drift and gradient flows, should follow curvilinear paths around the isobaths of the subtropical mound. Owing to the persistent meridional influx of water, the equatorial current becomes stronger as it flows away from the eastern boundary.

Since the ocean is also bounded at the west, high-pressure areas are formed in the south-west corner, where the water is piling up. Thus a thin western flow, driven by pressure gradient, should exist to replace losses of water on the northern side of the gyre due to Sverdrup transport.

This simple explanation for the asymmetry of a subtropical gyre ignores a lot of important details,

but the reasoning is consistent with the roles that rotation and friction play in balancing the pressure gradients. However, within this framework of momentum balance it would be difficult to explain why the Gulf Stream is so narrow and fast. Some more insight into this phenomenon is provided by vorticity balance in the ocean.

5.3 Vorticity theory of the Gulf Stream on the β-plane

Stommel (1948) published an article entitled 'The westward intensification of wind-driven ocean currents', which turned out to be a milestone in dynamic oceanography. Using a minimal amount of experimental data, namely, information on the zonal (along the parallels) distribution of wind velocity components over the Atlantic Ocean (a

6000 × 10 000 km rectangle†), Stommel successfully explained one of the most important phenomena of oceanic crculation. For the sake of simplicity, he assumed that only easterlies exist in the tropics and only westerlies at moderate latitudes (30°–40°N).

The first approximation used by Stommel concerned the ocean's shape. In order to retain a rectanrular co-ordinate system as a frame of reference, the spherical surface was 'unbent' so that the ocean became planar. The variations in spinning of various latitudes were considered by using the assumption that the Coriolis parameter $f = 2\omega \sin \phi$ varies along a meridian directly proportional to the distance from the equator, y (rather than the sine of the latitude, i.e. $f = 2\omega \sin \phi$)

$$f = f_0 + \beta y \qquad (5.1)$$

where β is a proportionality coefficient and f_0 is the value of f at some point where the ocean's plane intersects the Earth's surface.

How serious was the error introduced by this idealization? With respect to the Coriolis parameter, this can be easily evaluated. From elementary geometry it is known that the length of an arc, dy, is equal to the product of the radius (in our case, the Earth's radius, R) and the angle in radians ($d\phi$) comprising the arc dy. Using the rules of calculus we have

$$\frac{df}{dy} = \frac{df}{d\phi}\frac{d\phi}{dy} = \frac{d(2\omega \sin \phi)}{d\phi} \cdot \frac{d\phi}{d(R\phi)}$$

$$(5.2)$$

$$= 2\omega \cos \phi \cdot \frac{1}{R} = \frac{2\omega \cos \phi}{R} \simeq \beta$$

Now it is easy to compute $\beta = 2 \times 10^{-13} \cos \phi$ $cm^{-1} s^{-1}$. Since for the entire Atlantic subtropical gyre $\cos \phi \simeq 1$, Equation 5.1 produces insignificant error. For instance, at $\phi = 40°$, the actual value of $f = 2 \times 7.3 \times 10^{-5} \sin 40° = 9.38$ s^{-1}; if we use Equation 5.1 with $f_0 = 0$ (the initial point is taken at the Equator), the value $f = 0 + 2 \times 10^{-13} \times 1.40 \times 60 \times 1.852 \times 10^5$ $cm^{-1} s^{-1}$ $cm = 8.88 \times 10^{-5}$ s^{-1} (the last two factors represent conversion of the arc in degrees into length, 1° of latitude being equal to 60 nautical miles = 60 × 1.852 × 10⁵ cm). Comparing the two above numbers, we see that the error is only 5%.

† Actually, it is the average of the Atlantic and Pacific subtropical gyres.

By shifting the β-plane – the common term for a planar ocean with a linear variation of the Coriolis parameter – back and forth along the meridian, it is possible to select values for f_0 and β such that the planar circulation will be insignificantly different from that in a spherical ocean.

Assuming that the motions develop on the β-plane, Stommel 'forced' the entire system (the rectangular ocean with zonal winds of sinusoidal transverse profile above) (Fig. 5.5) to rotate with an angular velocity that increased linearly in the northward direction. He further assumed that the ocean floor is flat and rather shallow (200 m) and bottom drag accounted for the friction term. With these simplifications the procedures described in Section 4.6 can be applied to obtain the vorticity equation (4.9), which allows computation of the transport stream function ψ. Since we cannot trace the entire solution, we will evaluate various terms of the vorticity budget in Stommel's model to explain the effect of westward intensification of currents.

Out of five terms in Equation 4.9, only three were retained by Stommel

$$\zeta_R + \zeta_T + \zeta_P = 0 \qquad (5.3)$$

We consider first the relative curl. It follows from Equation 4.6 that the magnitude of ζ_R largely described the intensity of the currents. For the sake of simplicity, let us compare ζ_R at the western and eastern coasts at the middle latitude of the ocean, where the total transport is directed along the meridian. Therefore

$$\zeta_R = \frac{\partial \overline{V}_y}{\partial x}$$

where $\overline{V}_y$ is the average y-component of velocity vector from surface to bottom. Furthermore, since the ocean is closed (the ψ function is nil at the coast), $\overline{V}_y$ increases with distance from the shore and is everywhere positive (in the sense of Figure 4.5).

Now we can use Equation 5.3 to evaluate the magnitude of ζ_R at both locations. The wind injects some negative vorticity ζ_T, which is constant throughout the ocean because the wind system is purely zonal and has a constant shear. When current flows northward (at the western edge) it picks up clockwise (negative) torque relative to the Earth. On the eastern margin, the current takes a southward track, and a positive vorticity is added.

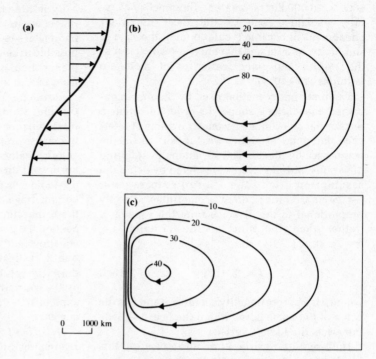

Figure 5.5 Stommel's model: sinusoidal wind distribution along the meridian (a) and stream functions in 10^6 m^3 s^{-1}. (b) Integral circulation in a non-rotating ($f = 0$) or uniformly rotating ocean; (c) integral circulation in an ocean rotating as the β-plane.

Compare the two balances (signs are shown in parentheses):

at the western edge

$$\xi_R^{(+)}\big|_W + \xi_\tau^{(-)} + \xi_p^{(-)} = 0$$

at the eastern edge

$$\xi_R^{(+)}\big|_E + \xi_\tau^{(-)} + \xi_p^{(+)} = 0$$

Now it is obvious that $\xi_R|_W \gg \xi_R|_E$, because $\xi_\tau^{(-)}$ is the same in both expressions. It means that

$$\frac{\partial \overline{V}_y}{\partial x}\bigg|_W \gg \frac{\partial \overline{V}_y}{\partial x}\bigg|_E$$

which suggests a very intense, narrow current just off the western coast opposed to a broad slow motion at the eastern boundary. Clearly, in a non-rotating ($f = 0$) or uniformly rotating ($f = $ constant) ocean, the relative vorticity should be equal to both shores, and the gyre will have perfect symmetry with respect to any axis crossing the center of the ocean. Stommel demonstrated these effects quite rigorously. His analytical solution is shown in Figure 5.5.

People looking for 'practical values' of models might be dissatisfied with this model, because it is too approximate (a rectangular closed ocean with a featureless bottom floor, wind blowing in only two directions, motion restricted only to the Ekman layer).

However, exploration of a complex natural phenomenon has never followed a straight, simple path. It is the extreme simplicity of this model that enables it to reveal a principal factor of global significance – the β-**effect**. Whereas all the other factors, such as variable depth and winds, non-uniform density and other variables, form major and minor peculiarities of the Gulf Stream system, the existence of westward intensification as such has now been explained.

Subsequent research focused on those factors and characteristics not covered by Stommel's model. The first major step in this direction was a model proposed by Munk (1950).

5.4 'Viscous' Gulf Stream

At first glance it might be thought that Munk (1950) added little to Stommel's model. Here again, a rectangular ocean rotating in the β-plane is used, although, in addition to subtropical gyres, his ocean includes the equatorial zone and the northern latitudes (up to 60°N). Only zonal wind

96

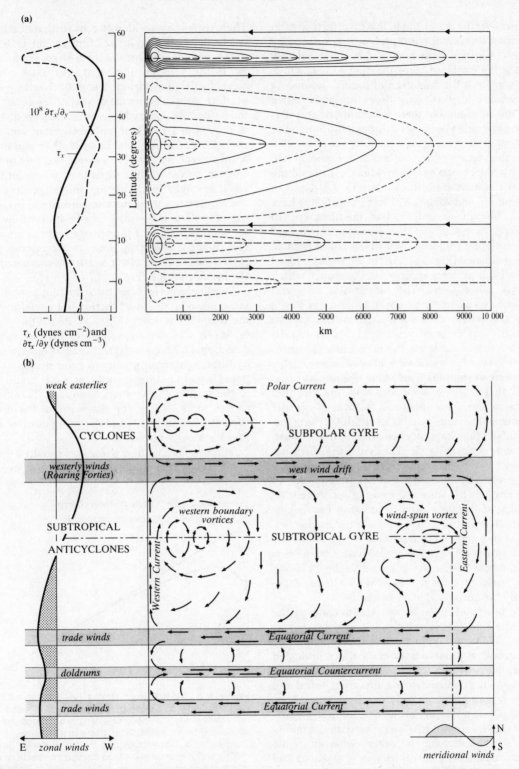

(a)

$10^8 \, \partial\tau_x/\partial y$

τ_x

Latitude (degrees)

τ_x (dynes cm^{-2}) and
$\partial\tau_x/\partial y$ (dynes cm^{-3})

km

(b)

weak easterlies

Polar Current

CYCLONES

SUBPOLAR GYRE

westerly winds
(Roaring Forties)

west wind drift

SUBTROPICAL

western boundary
vortices

wind-spun vortex

ANTICYCLONES

SUBTROPICAL GYRE

Western Current

Eastern Current

trade winds

Equatorial Current

doldrums

Equatorial Countercurrent

trade winds

Equatorial Current

E   zonal winds   W

N
S

meridional winds

Figure 5.6 (a) Streamlines of total transport and (b) schematic representation of the integral circulation driven by zonal winds on a rectangular non-uniformly rotating planar ocean. The meridional distribution of the wind stress τ_x and wind-stress curl are shown top left. A schematic representation of the wind system is shown below left.

blows over the ocean, but in this case, instead of an abstract sinusoidal wind profile, it is defined by the averaged distribution of the zonal velocity along the meridian corresponding to real winds blowing over the Atlantic and Pacific Oceans. As in Stommel's model, only three curls are balanced by the equation, but there is an important difference: viscosity is expressed through the horizontal exchange coefficient A_h described above (p. 000) and the average eddy gradient as in Equation 3.9. Using Shtockman's method, Munk calculated the total transport from the surface to the depth of no-motion and assumed it to be constant. Munk thus introduced only a few modifications to Stommel's model, but they were important, and the resulting picture of total circulation became quite different (Fig. 5.6). First of all, several closed circulation systems appeared in the model, which Munk called gyres. The latitudinal, east–west boundaries of the gyres moved to latitudes where the tangential wind stress was maximum and $\text{curl}_z \tau = 0$ (for a zonal wind, the curl is determined by the gradient τ along the meridian). The east–west axes of the gyres were situated where $\text{curl}_z \tau$ is maximum and the wind stress changes its sign.

With a degree of approximation, the resulting scheme represents some real features of oceanic currents. The large gyre in the middle of the ocean simulates the subtropical circulation in the North Atlantic or North Pacific. The gyre to the south of it resembles the system of currents near the Equator – for equatorial currents and countercurrents. In the north, in the zone of the westerlies, there is an analog of the North Atlantic or North Pacific current. Noteworthily, the gyres are not connected. They are separated by the line $\psi = 0$ as though by an impermeable vertical wall. Munk's model was the first comprehensive quantitative theory allowing comparison of the individual effects, especially the horizontal eddy viscosity.

At this point it would be useful to compare the Stommel and Munk solutions. A circulatory picture, similar to Figure 5.6, could be obtained by Stommel's approach on the basis of a complicated meridional wind profile. However, the difference between the two models is quite pronounced and it is related to the underlying physical mechanisms. Close proximity of the bottom in the Stommel model imposes frictional retardation, braking the flow throughout the ocean, whereas in the Munk model only lateral friction is essential and manifests itself only within the boundary layer, but is largely inessential in the rest of the basin.

Thus, two regions with different vorticity balance can be distinguished in the Munk model. Outside the immediate influence of the western boundary, i.e. in the 'interior', the wind-stress curl is balanced by the planetary curl, and the Sverdrup solution is valid over the major part of the ocean. In the western boundary layer, which has about 1/13 of the ocean's width, the planetary curl and the curl of lateral stress balance each other, and local wind stress does not play a significant role in the western current. This means that the amount of water involved in circulation solely depends on the wind stress in the interior, and lateral friction can only affect the width of the boundary flow. In the Stommel model the vorticity regimes are not that different and streamlines in the interior are spread more evenly along a latitude compared to the Munk solution (Fig. 5.7).

Another important feature of the Munk solution is the behavior of the ψ function in the vicinity of the western boundary (Fig. 5.7). Recall that $S_y = \partial\psi/\partial x$ (Eq. 4.10), i.e. the value and direction of the meridional (along y-axis) transport is determined by the longitudinal gradient of the stream function. Hence, a narrow countercurrent is predicted to the east of the western boundary. This flow carries some 17% of the discharge of the main current. This finding is an interesting result of the Munk viscosity model.

Some oceanographers expressed certain doubts that the above counterflow reflects the real world.

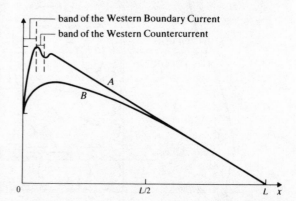

Figure 5.7 The transport stream function, normalized with respect to Sverdrup transport and divided by (sin ny)/M (M is the meridional length of the ocean), is shown for Munk's solution with lateral friction (curve A) and Stommel's solution with bottom friction (curve B). Stommel's solution shows decreased transport because of the effect of friction in the basin. Munk's solution oscillated near the western boundary, giving rise to a weak countercurrent (II) to the east of the main northward flow.

It is not obvious why purely viscous boundary layers should cause some water to return. Later studies showed that similar effects can be simulated in various non-viscous models. In any event, direct measurements performed by GEK (Section 2.5) and geostrophic calculations (Figure 5.11 below was drawn using such data) suggest that counterflow exists. This effect became important recently because many of the Gulf Stream rings are carried southward by this flow. The eastward continuation of the Gulf Stream and its southward countercurrent to the east of the main flow received a special name, 'the recirculation region'. Many efforts to explain and model physical processes leading to recirculation east of the Gulf Stream have been undertaken in the past two decades. Some of them we are going to discuss.

5.5   Gulf Stream as an inertial flow

It has been shown before that the total transports of the western current and the 'viscous' countercurrent in the Munk model are not affected by lateral stresses, but the latter largely determine the width of both flows. Using a value of A_h of 5.0×10^7 cm^2 s^{-1}, Munk obtained the corresponding width of the Gulf Stream of about 200 km. More recent studies of the Gulf Stream indicate the cross-sectional width of the main current to be one-third to one-fourth of the value obtained by Munk. Then the corresponding value of A_h should be reduced to about 10^6 cm^2 s^{-1}. Such an adjustment of A_h is frequently used in oceanography, since the eddy coefficients are not physical constants but largely depend on current structures. However, with relatively weak lateral viscosity, the role of inertial forces in the momentum balance may increase.

So far we have been discussing models of steady-state currents, disregarding acceleration or the force of inertia. However, acceleration of a fluid particle in a continuous medium consists of two components: a local variation that would be observed even in a totally uniform flow (such as the general deceleration of the current), and an advective component of acceleration created when a particle migrates to a part of the stream moving with a different velocity. This second type of acceleration will certainly occur even in a steady-state current as long as its velocity has some variation in space. A simple example of advective acceleration is the jolt experienced by a brash pedestrian jumping onto a trolley car moving at a constant speed.

We start our investigation into the nature of inertially driven boundary flows from several important approaches concerning the conservation of vorticity in a frictionless ocean. All these ideas are based on the principle of conservation of potential vorticity introduced into geophysical fluid dynamics by C. G. Rossby in 1940.

We, as inhabitants of Earth, can observe gyration only relative to the Earth, the relative vorticity ζ_R. However, actual vorticity, which would appear to an outside observer, consists of relative and planetary vorticity, $\zeta_p = f$ (analogy with the arguments revealing the Coriolis effect (Section 3.3) would be useful at this point). The sum of these two vorticities, $\zeta_R + f$, is called **absolute vorticity**.

Consider a column of fluid of depth h whose relative vorticity can be adequately described by the vertically averaged velocity vector. Then, in the absence of friction, changes of absolute vorticity in the selected column can occur only if this column exchanges its water with the ambient environment. When water converges into the column, its angular rotation, and hence absolute vorticity, increases, while diversion (outward horizontal motion) decreases absolute vorticity. These effects are absolutely similar to the spinning of a ballet dancer, who stretches or retracts her arms to control her speed of rotation (Fig. 5.8).

Intuitively, using considerations of continuity, one can conclude that the above horizontal water exchange should be related to changes in the average vertical velocity component $\overline{V}_z$ along the vertical, z. Therefore, the relative (to the total) time changes of absolute vorticity can be written in the form

$$\frac{d}{dt}(\zeta_R + f) = (\zeta_R + f)\frac{\partial \overline{V}_z}{\partial z} \qquad (5.4)$$

However, the last factor is simply the time changes of the vertical stretching of a water column of height h, divided by that height:

$$\frac{\partial \overline{V}_z}{\partial z} = \frac{1}{h}\frac{dh}{dt}$$

Upon substitution into Equation 5.4 and integration we obtain

$$\frac{\zeta_R + f}{h} = \text{constant} = F(\psi) \qquad (5.5)$$

Allegro　　　　**Andante cantabile**

Figure 5.8 The ballet dancer presses her arms along her body ('convergence') to increase torque, while angular momentum is conserved. The opposite movement, stretching arms and a leg ('divergence'), results in a decrease of angular velocity.

The quantity on the left is called the **potential vorticity** and Equation 5.5 represents the principle of conservation of potential vorticity. The constant on the right-hand side may differ from trajectory to trajectory. Since in steady-state motion the trajectories and streamlines coincide (see discussion at the beginning of Section 2.5), the potential vorticity is conserved along a streamline ψ, and the constant in Equation 5.5 is actually a function F of the streamline ψ.

Equation 5.5 illustrates the effect of vortex stretching due to its displacement along the meridian (changing f). Changes in height h of the column (topographic effect) also affects the relative vorticity ζ_R if the rotating column remains at the same latitude (f is constant). In the Northern Hemisphere vertical stretching is associated with inward motion (Fig. 5.9a) which, upon deflection by the Coriolis force, produces a positive relative vorticity (counterclockwise rotation). Shrinking of the column (Fig. 5.9b) causes outflowing motion which is deflected so as to produce clockwise rotation and, consequently, a negative relative vorticity. The values of h and ζ_R change simultaneously, so that overall the potential vorticity will be conserved.

Now assume that the height of the column remains constant, while the column moves meri-

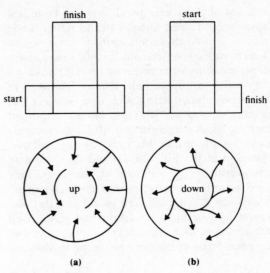

(a)　　　　**(b)**

Figure 5.9 Production of relative vorticity divergence and convergence of fluid columns: (a) influx of water induces positive vorticity in the Northern Hemisphere; (b) outflux of water causes negative vorticity.

dionally polarward. Since f increases, a negative vorticity ζ_R is developed to keep the potential vorticity constant. Similar considerations indicate that an equatorward flow of constant depth will acquire positive vorticity. These effects explain the circulation in a model developed by N. Fofonoff (1954), which was one of the first theoretic approaches to inertially driven ocean flows.

Consider a rectangular frictionless ocean rotating as a β-plane. It is easy to perceive that a steady-state motion should exist in such a basin since a linearly increasing moment of inertia toward the northern boundary will induce clockwise rotation. The above considerations suggest that northward flow will pick up negative vorticity which causes increasing crowding of streamlines down the flow, while at the western boundary the streamlines will be gradually diffused in the equatorward flow. Since vorticity remains constant in zonal flows, because f and h are constant, the model yields a uniform, westward flow in the interior, two symmetric boundary layers with northward flow at the west, southward flow at the east, and a jet across the northern edge (Fig. 5.10).

Although Fofonoff's solution appears to be very artificial, it demonstrates that purely inertially driven circulations are possible in the ocean, though the source of the moment might be different from irregular rotation. This model had an impact on later studies aimed at investigating the

100

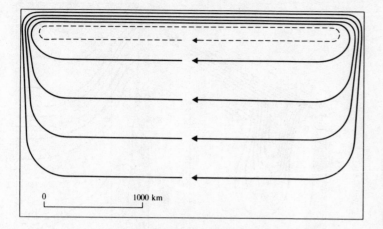

0   1000 km

Figure 5.10 Fofonoff's (1954) inertial flow pattern for a non-uniformly rotating ocean of constant depth. An inertial boundary layer diverts the steady westward flow in the interior to the north and an eastward jet is formed. The latter feeds into an inertial boundary layer on the east that supplies the steady westward flow of the interior.

role of inertia in the Gulf Stream and in the recirculation region.

If one attempts to continue the logic of the preceding arguments and trace the responses of relative vorticity to simultaneous variations in f and h, the effect will not be immediately obvious. It appears even more difficult to translate the above results of actual conditions in the western boundary region.

It had become obvious by the early 1950s that the Gulf Stream is so narrow that it cannot be balanced only by viscous forces. It looked like an intense jet of high inertia. But how can inertial forces maintain the flow over thousands of kilometers?

After 'futile attempts to construct to theory of an inertially governed Gulf Stream', in 1955 Stommel posed this problem to two experts in the field of non-linear geostrophic flows, J. G. Charney and G. W. Morgan. He also shared with them some ideas on the inertial nature of the Gulf Stream, based on available observations, which both scientists used independently to construct similar models. It is worthwhile to discuss Stommel's arguments first.

Simple considerations suggest that the effects of irregular bottom topography should be immaterial in the case of the real Gulf Stream, because noticeable velocities are not observed that deep. The major motions are concentrated above the main thermocline (Fig. 3.12), which constitutes a certain 'liquid bottom' for ocean flows. One might expect that in areas of high transverse velocity shear the relative vorticity and depth of the thermocline should be adjusted to each other so as to ensure conservation of potential vorticity along

the flow. This effect must be particularly pronounced in areas of sharp vertical density variations ('the discontinuity layer').

The Gulf Stream region is one such area. In Figure 5.11 the current profile and density (thermal) structure are portrayed in such a way that the topography of the thermocline (it can be identified by the 10°C isotherm) can be compared with shear effects. Let us focus attention on the inflection point of the 10°C isotherm. It is located in a plane which roughly coincides with the axis of the Gulf Stream. The isotherms are arched upward to the east and downward to the west of the flow. The relative curl has different signs on either side of the stream. Since the current is directed northward, the vortex stretching should change from west to east of the flow. On the western side the positive vorticity should be damped by planetary curl, whereas on the eastern side the negative vorticity increases northward. As a result, the slope of the thermocline should increase in the northward direction. This effect can be traced in the real ocean. Stommel mentioned these observations to Charney and Morgan to elucidate the phenomena to be modeled. Also the sharp discontinuity between the upper and lower layers was used to approximate the real continuous stratification by a two-layer model with geostrophic balance in the upper layer.

Again, as before, the ocean in both models was considered as a rectangular basin, rotating as a β-plane. But now special attention was focused on the narrow western boundary layer, where the effects of local winds were negated. Transport into the boundary region was assumed to be known from the Sverdrup balance in the interior or from

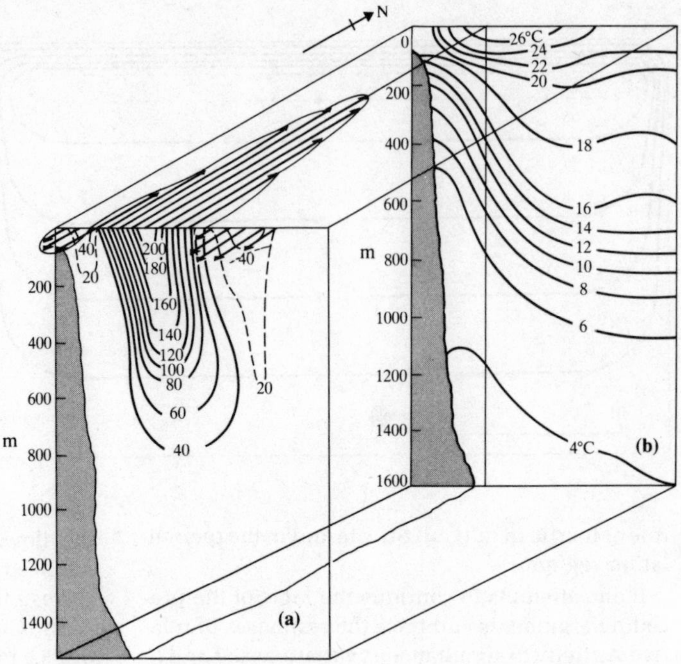

Figure 5.11 Block diagram of currents and temperature distribution in the Gulf Stream. Isovels of current are shown in plane A, isotherms are plotted in plane B. The vectors of average velocities are shown in the horizontal plane (the Gulf Stream velocities are positive, counter-current is negative). The Gulf Stream axis coincides with the vertical separating the zones of different curvatures of isotherms. (Three-dimensional compositions of diagrams in figures 7, 4 and 10 in Worthington 1954. Reproduced with permission.)

observations. Boundary flow develops in the upper layer, whereas the deep layer is assumed immobile.

Since the functions for the average velocity $\overline{V}$ and the depth h are unknown, the equation for potential vorticity (Eq. 5.5) was supplemented by another equation in both models

$$\frac{\overline{V}_y^2}{2} + \frac{\Delta\rho}{\rho} gh = G(\psi) \qquad (5.6)$$

where $\Delta\rho$ is the density difference between the two layers and ρ is the density in the deep layer. This is the Bernoulli equation in its simplest form, which accounts for energy conservation: the sum of kinetic (first term) and potential energies are constant along a streamline ψ. Arbitrary functions $G(\psi)$ and $F(\psi)$, which are constant along each streamline, are found from the following physical considerations. The relative vorticity ζ_R and velocity V_y are essential only in the western boundary layer. In the interior, where flows are diffuse and slow, the first terms in Equations 5.5 and 5.6 become negligibly small compared to the second terms. Therefore, the functions $F(\psi)$ and $G(\psi)$ can be found from the known depth h and vertical density difference $\Delta\rho$. Of course, the patterns of streamlines at the boundary, where the arbitrary functions $F(\psi)$ and $G(\psi)$ were fixed, should be known in advance.

Charney used existing observations, whereas Morgan simulated flows in the interior region.

Charney's results are illustrated by a block diagram (Fig. 5.12). The slope of the interface between the two layers increases along the inertial flow, which attracts some water from aside. Finally, inertial forces cause the interface to rise to the surface at a latitude corresponding to Cape Hatteras. Charney's model could not extend beyond this point.

The Morgan model was more general. It included a separate solution for the interior, a formation region for the boundary current (which he analyzed using the same model as Charney), and a northern region, where the Gulf Stream starts to 'decay'. He was one of the first to indicate that the torque exerted by the wind in mid-ocean should eventually be balanced by the bottom and lateral vorticity effects at the northern boundary.

The purely inertial theories of the Gulf Stream simulated two important mechanisms of western intensification. They explained at least why the stream is narrow and getting thinner northward, but only the main flow was treated in the above studies. In the 1960s extensive efforts were devoted to investigating the return flows and eastward jet-like continuation of the Gulf Stream forming the recirculation region.

Among others, G. Veronis (1966) managed

102

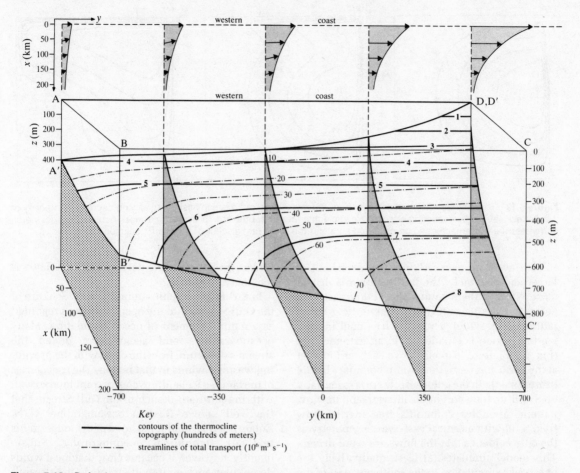

Figure 5.12 Defant's perspective representation of Charney's model displays the topography of the thermocline and the streamlines of total transport near the western boundary. At the top are shown the calculated velocity profiles across the Gulf Stream for several cross sections.

to elucidate that adherence of the southward countercurrent largely depends on the relative importance of the inertial and viscous forces or the inertial and Coriolis forces. The ratios between these pairs are characterized by two non-dimensional numbers, widely used in geophysical fluid dynamics. The first is the Reynolds number Re (Section 3.6), known to the reader, though now the horizontal eddy coefficient A_h should be used in Re instead of the molecular viscosity μ. The second parameter is the so-called Rossby number, $Ro = V/lf$, representing the ratio of inertial effects V/l (l is the horizontal scale of the flow) and the Coriolis parameter f. Several studies done by Moore (1963) and separately by Ilyin and Kamenkovich (1964) demonstrated that if $Ro \ll 1$ or $Re < 5$ (inertial effects are relatively small compared to friction and Coriolis forces) the circulation driven by zonal winds of Figure 5.5 tends

toward the Munk pattern. With larger Ro or Re the oscillations in the upper curve of Figure 5.7 extend to the east and eventually fill the entire northern half of the basin. These effects bear some realistic features of oceanic flows, though the observed recirculation in the north-west corner is missing.

Veronis discussed stronger inertial effects using the Stommel model with bottom friction. When Ro increases, the effect of 'overshooting' (as Veronis termed it) occurs in the model. This means that inertia in the western current intensifies in the north so that a fluid column overshoots the north-ernmost latitude that it had in the interior. Hence, a new boundary layer region must be generated (offshore of the original one) where friction and inertia drive the fluid column southward to its original latitude. In Figure 5.13a the return flow is stronger close to the boundary layer than it is farther to the east.

103

(a)
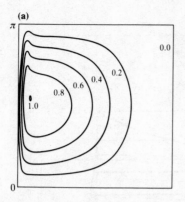

(b)
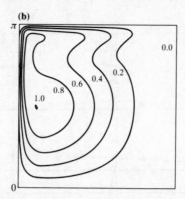

(c)
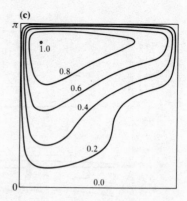

Figure 5.13 Three stream function patterns from Veronis (1966) for an ocean basin with varying degrees of intensity of wind stress. (a) The overshooting effect of a particle that returns southward. (b) A much stronger inertial effect. (c) Inertia dominates the system, creating an eastward jet along the north reminiscent of Fofonoff's solution.

With even stronger driving, the overshoot is larger and eventually the fluid column is driven close to the northern boundary and then eastward before it starts its southward track to its original latitude (Fig. 5.13b). Thus, the frictional inertial western region is broadened. In an extreme case (Fig. 5.13c), fluid 'parcels' move eastward in a jet at the north and reach the eastern boundary before turning south. In the latter case, there is essentially no Sverdrup transport in the interior, and the flow pattern resembles Fofonoff's free inertial flow (Fig. 5.10) with a central east–west asymmetry as the only evidence that the flows are wind driven. This model simulates, at least, qualitatively the observed recirculation to the south and east of the Gulf Stream after it has separated from the coast.

5.6 Why does the Gulf Stream leave the coast?

In the inertial models discussed so far, the Gulf Stream is deflected by a northern wall that does not exist in reality. The viscous models of Stommel and Munk predict an intensifying current along the western boundary only to the latitude of maximum wind-stress curl (approximately 33°N in Fig. 5.6a). Polarward of this maximum, it is not the Gulf Stream observed in the real ocean but rather a broad slow flow which returns into the ocean interior along the latitude of the maximum wind stress. The width of this flow is determined by the meridional scale of the wind-stress curl. Thus, the mechanism of separation of the Gulf

Stream from the continental slope at Cape Hatteras remains ambiguous.

In earlier geographic studies, the sudden turn of the Gulf Stream into the open ocean was regarded as a natural element of ocean circulation. Many oceanographers and geographers linked the stream's departure from the coast with the prevailing westerly winds in that region. The trajectories of the famous Icelandic cyclones coincided so well with the northern branch of the Gulf Stream that the well known Russian oceanographer N. N. Zubov (1947) called it the 'road of cyclones' in his textbook *Dynamical oceanography*. Subsequently, however, it became clear that local winds do not play such an important role in the stream's dynamics. It was discovered, in fact, that the Gulf Stream leaves the coast far south of the area of prevailing westerlies. In the open ocean the Gulf Stream remains as narrow as it was near the coast, with intensive meandering uncorrelated with the wind. The majority of Gulf Stream rings are formed just in this area. And yet, in spite of these arguments, the wind stress turned out to be an important force in the separation mechanism.

There were several unsuccessful attempts in the 1960s to explain the departure of the Gulf Stream from the coast using models of a homogeneous ocean driven by zonal winds. Quite a satisfactory explanation was given by Parsons (1969) and then independently by Veronis (1973). The latter study is very convincing and the arguments are followed below.

Consider a two-layer rectangular ocean driven as before by zonal winds of different directions (trades and westerlies), but now the meridional dimensions of the ocean are larger than the wind system. Each layer has constant but different density, and there is no stress at the interface. Inertial effects are assumed to be negligibly small ($Ro \ll 1$).

The meridional flow in the interior is a combination of geostrophically balanced motion and Ekman drift. If the flow were completely geostrophic, the total transport in the basin would vanish and the depths of the interface at the eastern and western edges would be equal ($h_W = h_E$). But Ekman drift, which is unrelated to the pressure gradient, accounts for part of the southward transport. Therefore, since the total transport is zero, there should be a net northward geostrophic transport, which will induce the west–east slope of the interface. This geostrophic flow is located in the west, so $h_W < h_E$. Thus, Ekman drift causes the thermocline (interface) to rise toward the surface. Separation of the Gulf Stream from the coast simply moves the western edge of the warm-water mass (upper layer) eastward so that the smaller Ekman drift acting on that water mass of more limited east–west width can just balance the geostrophic flow determined by the slope of the thermocline.

This result is similar to the circulation process discussed in Section 4.8 for a shallow sea or lake, where the Coriolis deflection was negligibly small. If a lake is thermally stratified, the wind blows the warm water to the leeward edge and causes the thermocline to rise on the windward side. The principal difference between the two phenomena is that the induced pressure gradient drives a vertical circulation in the lake, whereas it is geostrophically balanced in the rotating ocean, thereby generating a horizontal cell. But the leeward piling up of water is the same in the two cases.

The same logic can be applied to trace the continuation of the Gulf Stream, if the northern wind system is considered. Eventually, the mid-ocean flow should reach the eastern coast, though an intense boundary current cannot be formed there. In the real ocean, the water of this current sinks and gives rise to deep circulation (see Ch. 7).

This theory using a steady, linear, quasi-geostrophic model misses certain details (e.g. the longitude of the separated current), but reveals the basic mechanism governing the phenomenon. The key elements of the process are the geostrophic balance of downstream velocity in the western boundary current, Ekman drift and a limited amount of upper-layer water.

5.7 The nature of the Gulf Stream meanders and rings

It has been mentioned already that the narrow Gulf Stream jet meanders in a peculiar horizontal pattern, making rings and undulations similar to a tortuous river. Some of the rings break away from the main flow, reminding one of the formation of ox-bow lakes or bayous from the meanders of a river. The meandering reaches its culmination north and east of Cape Hatteras, where the jet makes a turn toward the open ocean. The rings and undulations, typically 100–400 km in length, move along the jet far slower than the average current velocity of the Gulf Stream.

Figure 5.2 illustrates the early stage of cyclonic (A) and anticyclonic (B) ring formation. In the former case, a large meander traps cold slope water in its 'pocket', and this pattern may be cut off from the main flow and move separately, rotating counterclockwise like the large eddy indicated by A'. Similarly, the warm core ring is about to be cast off, rotating clockwise as does its predecessor B'. The number of cold core rings is usually larger, e.g. 8–14 compared to the 3–4 that can be found north of the Gulf Stream. Some of the rings can be traced for months until they finally lose their identity or coalesce with the main stream. Their mean lifetime is estimated at 1–1.5 years, with a probable maximum of 3 years (!). The surface speed of rotation can reach 150 cm s^{-1} (3 knots) at the time of formation. Similar effects have been observed in the Kuroshio system.

Although the existence of rings has been known for almost 45 years, it is only in the last decade that their behavior has been understood in some detail. However, the theoretic hypothesis of ring formation was formed earlier. To begin with, it is worthwhile to take a look at the earlier views on the mechanisms of meanders and rings, and then to pass on to the more recent ideas.

Is there a theoretic explanation for the Gulf Stream breaking up into eddies and meanders? What kind of perturbations are these? What prevails here – horizontal turbulence or some other disturbance of a larger scale? In Chapter 3, we mentioned that the energy of the main current is dissipated by horizontal eddies. The narrow current widens and the velocity on its axis diminishes, but the general direction of the flow should remain unchanged. In the meandering zone, however, we observe this surprising fluctuation of the entire flow. The source of instability, in time, might have a larger spatial scale than the flow itself. To account for this, we must step out of the theoretic framework presented so far, for the phenomena involved are not embraced by the previously described mechanisms of fluid dynamics.

Theoretic studies of meanders do not consider eddies of a smaller scale (10–60 km), which are dissipated gradually and can be regarded as elements of horizontal turbulence. They are mostly concerned with those stream fluctuations whose amplitudes grow with time up to the point of the breaking away of parts of the jet and the formation of moving meanders.

Presently, there is no consensus concerning the nature of the meanders. One of the earlier concepts attributed them to waves of a special kind. The wave nature of the disturbances in the flow can be understood from the following simple scheme (Fig. 5.14). Two uniform flows move one above the other at constant but different speeds. The top layer has a lower density than the bottom layer. At the interface of the flows, waves will arise and, at a certain velocity gradient, with their amplitude growing, they may become unstable.

If the interface between the two layers is tilted or vertical, the waves propagating in the horizontal plane may, upon achieving a certain amplitude, break off a rotating mass of water of a certain density and propel it into the area occupied by the other water mass. Incidentally, cyclones are produced in this fashion – as wave disturbances of atmospheric fronts at the interface of air masses with different temperatures and wind velocities.

Applying the wave theory to the Gulf Stream seems legitimate. Indeed, here we have distinctly sloping boundaries between the upper and lower layers, emerging to the sea surface near Cape Hatteras. The Gulf Stream is treated as the main flow, with superimposed meanders representing the disturbances. Actually, wave-like patterns can be observed in the infra-red imagery of the Gulf Stream downstream from the Carolina coast (Fig. 5.1). Evaluating the amplitudes of waves of different length, Stommel (1958) showed theoretically that unstable waves can arise in the Gulf Stream at the observed current velocities. The length of these waves should be close to 180 km, which matches the size of the meanders. There was one perplexing circumstance, however: the theory describing the process is only applicable to perturbations of a small amplitude compared to the transverse dimension of the stream. Therefore, in this concept the Gulf Stream width had to be assumed to be many times larger than its real value.

Stommel (1958) attempted to elucidate the cause of the perturbations responsible for the meandering of the current during the course of observations conducted in June, 1950. He hypothesized that the process may be triggered by wind. In effect, in a certain segment of the current, the flow could have been reduced by a northerly wind creating a temporary obstacle by surface friction. But, whether the meanders originated from this obstacle or were due to some other cause was anybody's guess. The meandering could equally have been due, for instance, to the influence of underwater mountains. It must be understood that phenomena such as the meanders and rings cannot be immediately related to a real local wind or other apparent disturbing factors.

The wave-like response of the Gulf Stream to some random instabilities does not appear to be the physical mechanism completely responsible for the process of meandering. The meanders and rings on the coastal margin of the flow can evolve in the manner described above. The sizes and velocities of the average rings on the eastern edge, as we will shortly see, show that the energy of these patterns is comparable with the energy of the main flow. We can attain a theoretic basis for the interaction between the rings and the main flow, a base different from wave-like processes, if we discuss the idea of 'negative viscosity' effectively used by V. Starr (1968) to explain mechanisms of various geophysical flows and astrophysical processes.

The essence of negative viscosity phenomena can be revealed by the following considerations. As we now know, the typical transfer of energy in oceanic currents occurs from larger circulation elements, the gyres, to smaller elements. In particular, oceanic turbulence is adequately interpreted as cascaded energy transfer from the average flow down to smaller turbulent eddies. In this case, the turbulent viscosity is expressed through the eddy viscosity coefficient by Equation 3.9. This relationship, and the entire mechanism of turbulence, provided a plausible description of the major properties of currents in oceans and seas. However, for jet-shaped currents, such as the Gulf Stream, this mechanism proved of little use. Its shortcomings were first seen in Munk's model, with its greatly overstated Gulf Stream width. In the inertial models of Charney and Morgan, with viscosity eliminated, the jet was narrower, but still not narrow enough. It became obvious that there

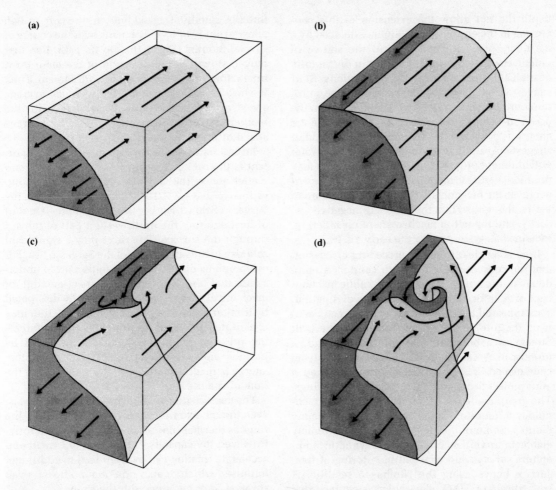

Figure 5.14 Formation of a meander or a cold-core ring from the horizontal undulation of flow. Originally, (a) two water masses (a lighter one superimposed on the denser) move in opposite directions. The denser water (shaded) may appear at the surface as a result of deflection of the upper flow (b). The internal instability causes a wave of small-scale amplitude (c), which may be amplified to a separating meander by deformation of the velocity field in both flows (d).

should be another process maintaining the jet in this narrow shape over a considerable length.

To arrive at Starr's idea, which at first seems paradoxical, let us call into question an assumption that, at first glance, seems indubitable – the primary nature of the jet and the secondary nature of the meanders. Starr suggested that, under certain conditions, the transfer of energy may occur from small irregular eddies to the larger average current. In this case, flows of energy would be directed from either side toward the axis of the stream, compressing it into a narrow jet. These energy flows are balanced by internal dissipation due to small-scale turbulence inside the jet. The turbulent exchange coefficient (or 'eddy viscosity') calculated for such a process proves to be negative, since the flow of momentum is directed

against the velocity gradient – hence the term 'negative viscosity'. For a model of motion with negative viscosity to be fully substantiated, it is necessary to determine a mechanism responsible for the influx of external energy into the irregular motions that transmit their energy to the average currents. However, the observational data available at that time were insufficient to define such energy sources with certitude.

Starr showed that the effect of negative viscosity was clearly observed in disturbances such as meanders. He calculated the magnitude and direction of the flow of kinetic energy of fluctuations by using the observed data obtained by GEK south of Cape Hatteras and near Florida. The resulting energy flow was found to be directed from the two sides toward the jet axis. Lack of data on currents at depth did not allow the extension of this conclusion to the entire water column embraced by a meander. Also, attempts to find the source of kinetic energy of irregular motions in the vicinity of the Gulf Stream ran into complications. Starr suggested that thermal convection might be one of the sources of energy. However, convection in the ocean develops too slowly to supply kinetic energy for the rapidly moving eddies. Another suggestion was that an intense mixing of water with different properties at the jet boundaries may produce vertical motions supplying sufficient energy to the meanders. This hypothesis is hard to verify, though, so that the source of negative viscosity, and the entire mechanism of meandering, remained a mystery up to the early 1970s.

It must be clear from the preceding discussion that the subsequent task was to conduct a more detailed study of the rings and the subtle mechanisms of their formation, evolution and dissipation. This appeared to be the only way to correctly estimate the role of the energy balance of the Gulf Stream as a boundary current. To this end, a number of American researchers formed a team code-named 'The Ring Group' and launched a joint project to investigate the Gulf Stream rings. The group made use of the most up-to-date technology to monitor the individual rings for many months. Some of the rings underwent extremely elaborate surveillance by means of synoptic temperature surveys and constant monitoring of free-surface buoys using the Nimbus 6 satellite. A sophisticated CTD-O_2 electronic system recorded vertical profiles of temperature, conductivity and dissolved oxygen from the sea surface down to the bottom. Even biological samples were taken at different depths to describe the distribution and migration of organisms within the rings. On top of that, infra-red satellite images (Fig. 5.1) allowed the easy identification and positioning of individual rings, which for convenience were given personal names, such as Al, Bob and Charlie. One of the rings, Bob by name, was studied in the greatest possible detail. Its infra-red image adjacent to the main flow is presented in Figure 2.11.

Bob became a detached ring after the main Gulf Stream took a big pocket-like meander and then a short-cut (Fig. 5.15). A large amount of cold, less saline slope water had been trapped in the ring, thus forming a peculiar bell-shaped distribution of isotherms (Fig. 5.15b) and other characteristics at the cross section AB. Bob was constantly monitored. A satellite-tracked buoy in the core of Bob allowed study of its movements over the course of several months (Fig. 5.16). On its path, Bob first moved slowly in the direction of the main flow, interacting sometimes with the Gulf Stream. Then it steadily moved south-west with an average speed of 5.5 cm s^{-1} until it coalesced with the mother stream in September, 1977. The average speed of the buoy was 125 cm s^{-1}.

From this example, we can gather how important is the role of rings in transfer of energy and matter across the Gulf Stream. The Ring Group estimated that 4.2×10^{22} J per year is lost from the Sargasso Sea, which is comparable to the amount of heat entering the northwestern part of the sea through the surface. The rings play a significant role also in the salt balance of the Sargasso Sea. For our previous discussion, it is important to understand that the rings carry a great amount of mechanical energy, which is not easily dissipated by friction. It has been estimated that friction may constitute only 1% of the principal driving force – the pressure gradient – which is balanced by Coriolis and centrifugal forces. Therefore, this energy is readily available to the main flow at the time of coalescence.

The above partially confirms Starr's hypothesis, even though external energy does not come to the rings as thermal convection nor in any other form. However, the capacity of the rings to extend and accelerate rotation can by itself lead to additional impulse, which makes the main stream even stronger after merging with the rings.

Successful attempts at modeling single eddies and the entire process of meandering were made by P. Rhines (1976), which showed that the formation of various eddies does not fit in with the deterministic framework of slow ocean models discussed so far. Another fast ocean model allows us to describe a combination of wave and steady-state motions, leading to the formation of a complex vorticity field. Models of this type cannot, of course, be used to make predictions. The mathematical equations governing eddy-contaminated ocean flows cannot be solved to forecast motion far into the future. Nevertheless, for an instance of prediction that is brief compared to the timescale of large flows, the computed picture can be fairly accurate. With each subsequent step, however, the error may grow, and a long-range forecast will produce a current field far from that obtained in actual measurements.

Let us return to the experiments just described.

(a)

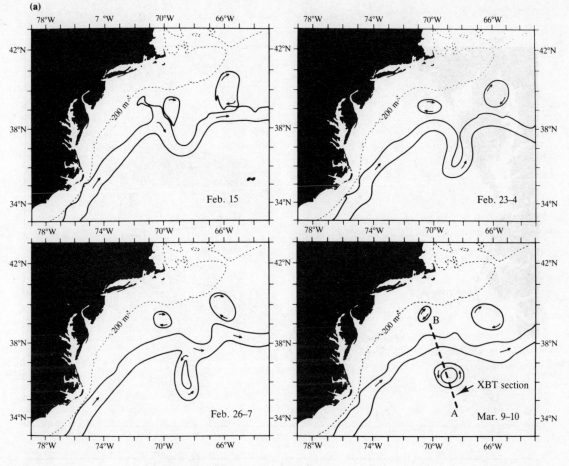

(b)

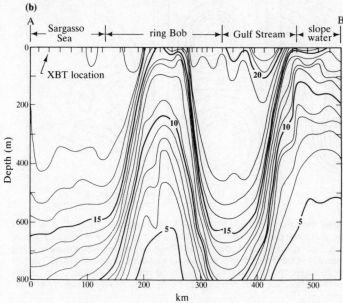

Figure 5.15 (a) Diagrams showing the formation of cyclonic Gulf Stream ring Bob in February–March, 1977, based on infra-red images from NOAA-5 satellite. Two anticyclonic (warm-core) rings were observed north of the Gulf Stream. (b) Vertical temperature section through ring Bob and the Gulf Stream on March 12, 1977. Ring Bob can be seen as the area of raised isotherms and cool surface temperatures.

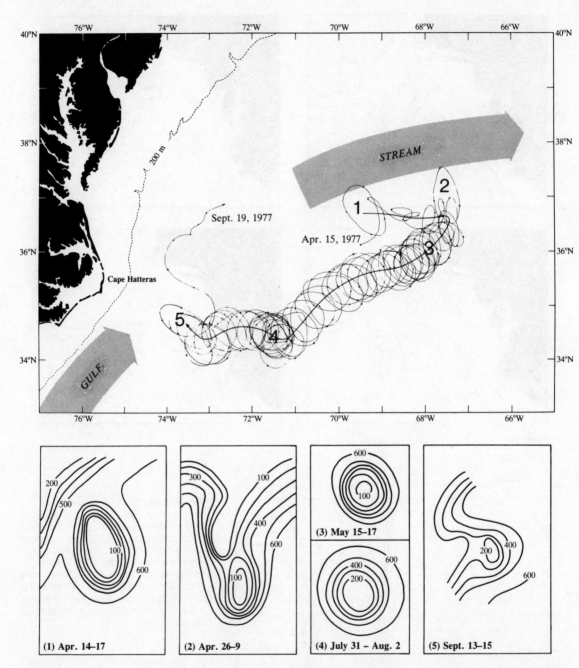

Figure 5.16 (a) Trajectory of a freely drifting buoy as it looped in ring Bob from April 15 to September 15, 1977. The mean position of the Gulf Stream is shown schematically. (b) Lower panels show depth contours (meters) of the 15°C isothermal surface in Bob at five times in the Bob's existence. During April, Bob became connected to the Gulf Stream and moved rapidly (up to 25 cm s⁻¹) eastward. In May, Bob separated from the stream and began its southwestward drift (5 cm s⁻¹) through the Sargasso Sea. In September, Bob rapidly coalesced with the Gulf Stream and was lost. The final coalescence may have been triggered by another ring, Dave, which advected downstream in the Gulf Stream into the vicinity of Bob.

The ring Bob exhibited very distinctive 'personality', unchanged for many months. A ring of this kind is a formation comparable in size to an inland sea, such as the Black Sea, which may take a modern ship about ten hours to cross. It is highly questionable whether we should try to describe mathematically and to predict the behavior of each ring or eddy. Meandering is triggered by a great number of purely accidental factors whose intensity and origins cannot be predicted. As in the case of turbulent eddies, we are concerned with the average shapes and sizes of the eddies and with the flow of energy from its source to the point where it is lost due to fluid friction. So, if numeric experiments such as those of Peter Rhines reproduce the general behavior of eddies, and the fluxes of energy and matter thus generated, they are already a great success. Major breakthroughs have been made along these lines. Without doubt, they were all made possible by data on the mechanisms of evolution and extinction of eddies obtained from numerous field experiments. Experimentation and theory go hand in hand in studies of the Gulf Stream, revealing one after another the hidden aspects of this tremendous natural phenomenon.

That we have discussed at such length the mysterious and intriguing properties of just one current was not without reason. In fact, the Gulf Stream attracted our attention because of its typical rather than its unique properties. We asserted, and then proved, as a general theoretic proposition, that intense currents in the west have a major part to play in the circulation of every ocean. Besides, the Gulf Stream happens to be one of the best-studied currents, certainly holding top place in the number of publications dealing with it. This latter fact has enabled us to review, with particular reference to the Gulf Stream, some of the major ideas in the theory of oceanic circulation, and to verify these ideas on observational data pertaining to the Gulf Stream itself.

The close attention paid by scientists to the various features of this current has revealed the internal mechanisms of processes observed in other oceanic regions as well; it has also turned the Gulf Stream into a testing ground for the improvement of the tools of experimental and theoretic oceanographic research.

But does the Gulf Stream as such possess any specific significance?

We opened our discussion on the Gulf Stream with the common but, it must be now admitted, incorrect statement that Europe's favorable climate is a direct consequence of the existence of this 'river in the ocean'. In actual fact, the relationship is neither direct nor simple. It would be more correct to say that the Gulf Stream as a whole (with its intensive thermal contrasts, meandering patterns, etc.), being a major component of the circulation of the entire subtropical zone of the ocean, serves as an indicator of this circulation. We have demonstrated, however, that the main cause of this circulation is the wind. Does this mean that the Gulf Stream is a consequence of the climate rather than vice versa?

Yes and no.

Stable, fair weather prevails over the calm, warm waters of the Sargasso Sea, which lies at the center of the subtropical gyre. This is the realm of the Azores atmospheric pressure maximum. Huge eddies – cyclones with low pressure in the center – travel along its northwestern boundary. The entire system – trade winds in the tropics and westerlies at moderate latitudes – is set in motion largely by the thermal energy that comes from the ocean.

Thus, the Gulf Stream, as part of the chain of energy transfer in combined air–sea circulation, is both cause and effect of climatic processes. Being the most important and powerful component of this gigantic whole, it is certainly worthy of the effort invested in studying it.

Finally, one more question should be answered to understand the westward intensification of currents.

5.8 Why are 'gulf streams' weak in the south?

Another paradox of oceanic circulation that remained unexplained for a long time is that the asymmetry of subtropical gyres, so obvious in the northern parts of the ocean, is less pronounced south of the Equator. Indeed, the Brazil Current does not seem much stronger than the Benguela Current, and the East Australia Current is even weaker than the Peru Current. The most powerful meridional current in the south, the Agulhas Current, also bears little resemblance to such jet-like currents as the Gulf Stream and Kuroshio. Passing near the southeastern coast of Africa, it does not turn into the open ocean, making a 'deflection', but penetrates further into the Atlantic, giving only a part of its waters to the Antarctic Circumpolar Current.

Why is there no westward intensification of currents in the Southern Hemisphere? As in the north, there are anticyclonic wind systems in the Southern Hemisphere producing the subtropical gyre, and the planetary vorticity should make it asymmetric. Since this does not happen, there should exist in the south some processes counteracting the β-effect.

The obvious cause of the weak western intensification might be some peculiar changes in the geometry of the basin. The polarward current may expand vertically, or horizontally, or both. G. Neumann (1956) studied the effect of vertical expansion of the current. He argued that the depth of the main thermocline increases polarward more rapidly south of the Equator than in the north. Actually, he considered the layer of water in which the upper currents are developed, e.g. the depth of no-motion. However, deep penetration of currents south of the Equator is not related to pure geostrophic balance. The powerful Antarctic Circumpolar Current and pre-polar convection ensure intense vertical mixing and appreciable currents at all depths. In the northern parts of the ocean, these dynamic processes are apparently weaker.

Using the Defant map of the depth of no-motion (analogous to the one in Fig. 3.16), Neumann showed that vortex stretching (Section 5.5) in the southward Brazil Current offsets planetary vorticity, and the relative vorticity does not increase polarward. He proved his conclusions by comparing the downstream distributions in the Brazil Current and the Gulf Stream. In the first case they are stretched vertically in a fan-like manner, suggesting noticeable vertical expansion.

The ocean thermocline and, moreover, the depth of no-motion are not, however, rigid surfaces, but largely depend on the intensity of the subtropical gyre. Therefore, the entire mechanism is rather ambiguous.

The later numeric models demonstrated that the geometry of the oceanic basins plays an important role in the intensity of subtropical gyres. Horizontally, all three oceans become wider southward and are open to the Antarctic circumpolar region, whereas the northern parts gradually taper off. The West Wind Drift in the south affects the entire subtropical gyre more effectively than do similar flows in the north. It is natural to assume that the intensity of the circumpolar motion determines not only flow characteristics in the southern part of the ocean but the thermohaline structure of the oceans, and even controls water exchange between the hemispheres. Assuming that the overall balance of heat and water flows (hence, average thermohaline structure) remains unchanged, it is reasonable to study the ocean's response to various

(a)

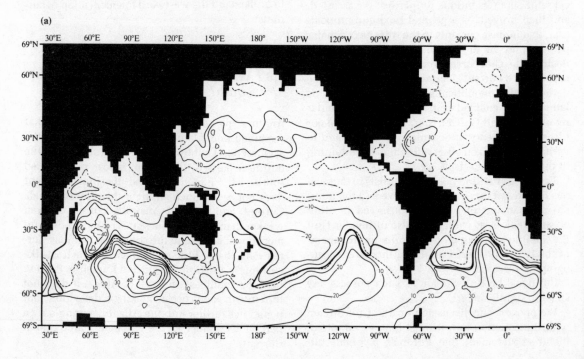

(b)

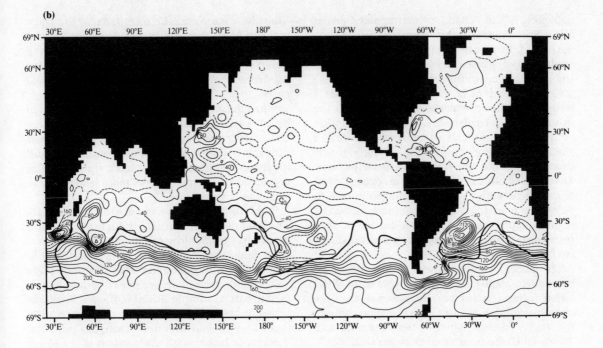

(c)

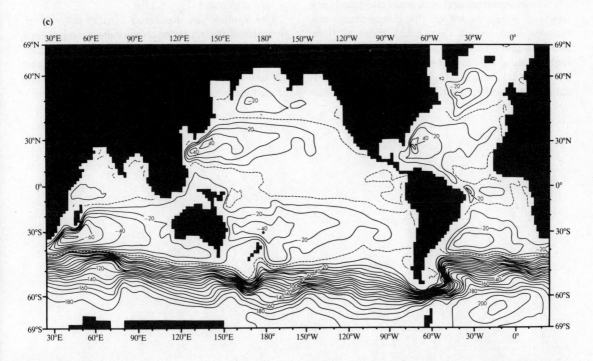

Figure 5.17 World Ocean circulation: (a) total mass transport stream functions for uniform density with realistic topography and $M = 22 \times 10^6$ ton s^{-1}; (b) stream functions in a diagnostic model based on the observed temperature and salinity fields with $M = 184 \times 10^6$ ton s^{-1}; (c) pattern of integral circulation for the predictive model based on a 2.3 year numeric calculation using the observed density and $M = 184 \times 10^6$ ton s^{-1}. M is the total mass transport by the Antarctic Circumpolar Current.

transports of the Antarctic Circumpolar Current alone. Such models, called **diagnostic models**, are now used frequently. Experiments with these models are, of course, only feasible using numeric simulation on powerful computers.

K. Bryan and M. D. Cox (1972) performed a series of numeric experiments simulating the overall oceanic circulation. A detailed description of their model would take up too much space and would require the introduction of certain new concepts that are irrelevant to oceanic physics. They carried out calculations for an ocean covering a sphere, with a resolution of 2° from 62°S to 62°N. The model, assumed to be non-stationary, requires approximately 2.3 years for the ocean's response to the initial conditions to become stabilized.

The model for uniform density with a realistic bottom topography (Fig. 5.17a) shows basic circulation patterns in the ocean. However, they are unrealistically weak, since mass transport through the Drake Passage between the Antarctic and Tierra del Fuego is assumed to be too small (22 × 10^6 tonne s^{-1}).

The subsequent experiment was conducted for a stratified ocean in which the temperature and salinity distributions were taken close to their annual average values. The mass transport of current around the Antarctic was increased to 184 × 10^6 tonne s^{-1} (Fig. 5.17b), but the current stabilization time was kept small. In this case the streamlines show more realistic transports for the western boundary currents in the Northern Hemisphere, although even so they are smaller than the observed values, the currents are broader and weaker, and there are a great number of small-scale gyres.

When the period of stabilization was assumed to be 2.3 years and the circumpolar transport was 184 × 10^6 tonne s^{-1} (Fig. 5.17c), the various small subgyres of Figure 5.17b almost disappeared, and the integrated transport was much simpler, being organized into primary gyres more familiar to us (Fig. 2.2).

One further feature is remarkable in Figure 5.17. The total potential vorticity in Figure 5.17a for a purely barotropic flow follows the lines of constant depth divided by sin ϕ. This is particularly pronounced outside the frictional boundary layer and the strong wind-forcing areas like the equatorial currents. A sample line of $H/\sin \phi$ is superimposed on the circulation patterns in Figure 5.17a (bold line). In the stratified ocean, the potential vorticity is linked with the position of the lower thermocline boundary, and the effect of the bottom is less important (Fig. 5.17b).

The studies just discussed display the role of thermohaline circulation, briefly described in Section 3.5. This process, which will be discussed in detail, largely controls the depth and sharpness of the main thermocline, and thus the vertical scale of surface circulation.

6 Countercurrents and eddies

6.1 Instabilities of ocean flows

In this analysis of large-scale ocean circulation, we have relied so much on the high mechanical and thermal inertia of the oceans that we have not bothered to look beyond very slow changing, actually stationary (steady-state) types of circulation. Any effects connected with variations on a scale smaller than the average flow pattern we grouped under the name 'turbulence' or 'turbulent eddies'. All energy and mass transport relations between the highly variable fields and average flows were grouped into a convenient quantity – eddy coefficient, which itself depends on the average flow field. The only assumption concerning the scales of turbulence was that of anisotropy, which implies that horizontal turbulent fluctuations are many times larger than vertical ones.

This scheme worked perfectly for the explanation of large-scale circulatory patterns in the ocean. One could imagine, though, that stationary motion would never occur in the real ocean. However, stationarity appears to be a very good approximation to the real world, because on a very long timescale the ocean responses to external forcing show low-frequency variability, and small-scale effects could easily adjust to each new mean field.

The difficulties began when the currents appeared to be narrow and swift, like the Gulf Stream. Its high inertia concentrated over a very small spatial scale was not balanced by the normal dissipating mechanisms. All kinds of new phenomena, like negative viscosity, overshooting and the formation of meanders and rings, appeared to be essential elements of the boundary flow dynamics. These effects clearly suggested that high-frequency variability dominates the average flow.

Let us recall the recirculation region east of the main Gulf Stream. We found that the southward counterflow was largely caused by overshooting due to the high inertia developed at the northern tip of the boundary current before it turned east. In addition, this region displayed the highest instability of the entire system, so the warm Gulf Stream rings are formed right there.

One may question how the overshooting is physically implemented. If the water parcels were elastic balls and the northern boundary were a solid wall, like that shown in Figure 5.13, the mechanism of equatorward motion could hardly cause any uncertainty. However, instead of a solid wall we have a sloping fluid boundary marked by high temperature gradients, and, instead of balls, rings largely mark the counterflow by invariably taking the southward track. They perhaps carry much of the energy in the recirculation region. These considerations bring us to the idea that counterflows and eddies generated by the main flow are highly interrelated.

Another important factor related to counterflows should not be overlooked. The machinery of the subtropical gyre has been explained without reference to the depth of the flow, i.e. adjustment of the density field was largely unnecessary. However, the effect of high inertia and departure of the boundary flow from the coast could not be successfully explained without the assumption that the lower boundary (thermocline) contributes to the vertical contraction and route of the jet.

The recirculation regions of the Gulf Stream and, evidently, the Kuroshio are the only ones of this kind in the World Ocean. Only there do planetary vorticity and friction (or inertia) make the boundary flow so intense and variable that it 'shakes off' a part of its energy via relatively organized rings which die away without losing their identity. Can we define stable ocean flows that are balanced by the 'old' energy-dissipating mechanisms described in Sections 4.1 and 4.2, i.e. by cascaded energy decay via random turbulent motions.

We do not expect to find such flows near the eastern coasts of the oceans, where currents are variable due to unstable winds and coastal upwellings. We will not be able to find very intense flows at high latitudes either because a high rate of vertical mixing due to convection would ensure large vertical scales of currents. Our attention should be attracted by the equatorial zone, where the surface layers are shallow due to an elevated thermocline (Fig. 3.12) and rather distinct flows and counterflows (Fig. 2.2) develop under very persistent wind systems (trades).

We will see that the mean currents around the Equator are stable and well defined, as follows from both the theoretical models and the distribution of properties. However, even the very first field experiments over large water areas using an advanced observational technique demonstrated that eddies with a scale larger than the mean flow are inherent in the motion. These variations, which sometimes acquire rotational structures, are hardly related either to the instability of the average flows or to direct wind forcing. They appear to be responses of some 'distant disturbances', which propagate in the ocean in the form of horizontal undulations or waves (called Rossby waves). Their behavior is heavily affected by the Earth's rotation and existing stratification.

It would be difficult to give a comprehensive picture of the interaction between average flows and eddies, because it is still a very active field of research. However, the following description is not a number of 'case studies' but rather an attempt to bring to the reader's attention what is known and can still be explained by our conceptual approach.

6.2 Features of the equatorial zone of the oceans

The tropical zone of the oceans has been at the focus of attention for a long time. The sea surface here receives the greatest amount of solar heat, which is then transferred by currents far beyond the zone, determining the climate and the weather over a large portion of the globe. Intensive transfer of kinetic energy from the atmosphere into the ocean occurs in the zone of the trade winds, the most stable of the Earth's wind systems. The South and North Equatorial Currents are parts (or branches) of the general ocean circulation.

The tropical zone has a remarkably complex system of currents and countercurrents. The sea water in this region has a no less complex structure.

Figure 6.1 shows a schematic profile of the Atlantic Ocean illustrating vertical transport and horizontal movement in the meridional plane, as well as a number of important oceanographic surfaces and a diagram of water motions. The diagram is a more elaborate version of Figure 3.12. At the center of a subtropical gyre, a convergence zone is

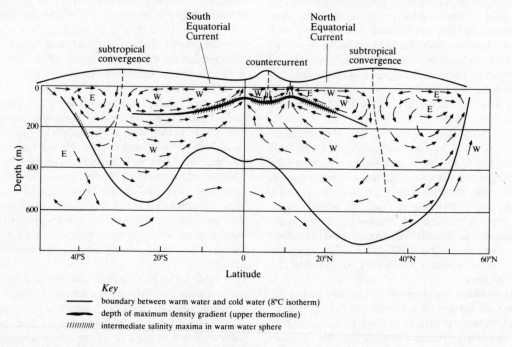

Figure 6.1 Schematic meridional cross section through the upper layers of the Atlantic Ocean. (After Defant 1936.)

116

formed. The zone of subtropical convergence is where the descent of water prevails. Part of the descending water goes toward the Equator, where it offsets the loss of warm water occurring in the surface layers. This dynamic mechanism of meridional circulation between moderate and tropical latitudes makes the cold deep water rise to the surface layers near the Equator. Temperature declines abruptly with depth, and the oceanic thermocline becomes more pronounced (these effects will be discussed in more detail in the next chapter). In addition to thermal factors, the mechanism of vertical movements is supported by an inhomogeneous evaporation–precipitation rate in the equatorial zone.

In the region of the trade winds, intensive evaporation increases salinity and hence density of the upper layers. This induces convection overturn, i.e., the sinking of dense saline water, and, at a depth of 100–200 m, a layer of elevated temperature and salinity is created (**tropical density discontinuity layer**). Just north of the Equator is the belt of prevailing calms or the 'doldrums'. This is the zone of frequent tropical rains. The rains decrease the salinity of the surface water, and vertical stratification is intensified, impeding convection. The tropical discontinuity layer is very sharp because it is formed by both temperature and salinity differences. It is actually an interface between two waters of different densities, which precludes momentum transfer downward due to turbulence. The top layer is almost homogeneous and does not mix with the underlying water, and therefore it is possible to apply the two-layer model with different densities and no mixing across the interface.

From a global point of view, the equatorial zone has an important feature. At the Equator, the Coriolis parameter is zero, and remains extremely small near the Equator. This means that the horizontal component of the Coriolis force disappears, and the geostrophic balance, largely controlling motions of the oceans and atmosphere at high and moderate latitudes, breaks down close to the Equator. As a result, elementary dynamics relationships should be different in the equatorial zone (within a few degrees in each hemisphere) than elsewhere.

This fact, together with the climatic features, determines the peculiar and complex kinematics of the near-equatorial waters.

6.3 The nature of equatorial countercurrents

Equatorial countercurrents were probably known back in the time of sailing ships. They are directed to the east, opposite to the North and South Equatorial Currents, crossing the oceans from the west coast to the east coast as narrow bands. In the Atlantic and Pacific Oceans, equatorial countercurrents are observed slightly north of the Equator in all seasons of the year. They are subject to fluctuations, achieving their greatest strength during summer in the Northern Hemisphere. This is particularly evident in the Atlantic Ocean. In winter, the countercurrent starts east of 24°W, whereas in summer it occupies almost the entire width of the ocean. In the Indian Ocean the countercurrent as an independent entity is observed only in winter, south of the Equator. In summer, during the southwestern monsoon, it is displaced northward and merges with the southwesterly current.

For a long time the cause of equatorial countercurrents remained unexplained. This phenomenon defied intuition, since it flows opposite to the prevailing winds. Montgomery and Palmén (1940) hypothesized that the easterlies (trades) in both hemispheres piled up water against the western coasts of the oceans, and thus that pressure gradient was the major cause of the equatorial countercurrent. This flow, they argued, should manifest itself in the doldrums, where the wind stress is reduced. They presented supporting observational evidence in the Atlantic.

However, in the Atlantic Ocean (Fig. 6.1) the countercurrent is located 5–10° to the north, and the Coriolis effect cannot be neglected, which means that wind-induced surface slope should have some transverse component due to Ekman transport. Another consideration is that, if the above hypothesis were true, countercurrents should be most strongly developed in the Atlantic and Indian Oceans, which have solid meridional boundaries, the continents. In reality, however, the most powerful countercurrent is observed in the Pacific Ocean, in spite of its being linked by wide passages to the Indian Ocean. The west–east pressure gradient is important, though, for an explanation of the undercurrents at the Equator, which is discussed in the next section.

The paradox of the equatorial zone was elucidated for the first time by V. Shtockman in 1945, who showed that a decisive role in the formation of the equatorial countercurrents is played by transverse wind inhomogeneity caused by the calm belt between the two trade winds. We already know Shtockman's ideas concerning the role of inhomogeneous wind in closed circulation in seas (See Ch. 4).

Theoretically, a narrow channel in the ocean is considered to stretch around the globe north of the Equator. The depth of this channel is assumed constant and conditions are formulated for its 'walls' to simulate free exchange with the environment. The Coriolis force is constant: the channel is involved in uniform rotation. A unidirectional (easterly) wind blows over the channel with its meridional velocity profile resembling the wind distribution over the area between the northern and southern trade winds (Fig. 6.2). The conditions of the model are zonally independent, and hence variations occur only across the channel. It is sufficient, therefore, to determine the currents in one cross section. Given uniform density, the problem in this formulation is reduced to Ekman's problem.

Figure 6.2 plots the system of streamlines and the free surface profile versus wind distribution. The zone of strong countercurrent is seen to pass through the area where the wind velocity is minimal. The reason for this is easy to see. Figure 6.2 shows a plan of the distribution of tangential wind stress along the y-axis (N–S lines). Since both trade winds blow in the Northern Hemisphere, Ekman transport should carry some water northward under each wind. The sea surface must be elevated at the boundary between the southern trade winds and the zone of the doldrums, and depressed at the northern edge of the doldrums (bottom part of Fig. 6.2). The countercurrent appears to be purely geostrophic. The south–north pressure gradient is balanced by the Coriolis force, and the steady-state flow goes eastward.

Later models took into account some important factors overlooked by Shtockman. The eastern boundary in the previously unbounded channel and linear variation of the Coriolis parameter f with latitude (instead of constant f) were introduced by Sverdrup and Reid. Their model largely repeats the Sverdrup model (Section 5.2) but it uses a more realistic meridional distribution of the wind stress. As a result, the streamlines representing the equatorial currents and the countercurrent show considerable exchange of water in the meridional direction due to Sverdrup transport,

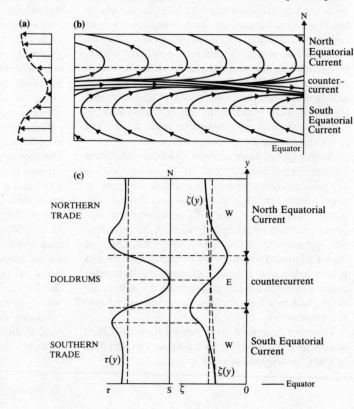

Figure 6.2 Distribution of streamlines in the tropical zone. (After Shtockman (1948b).) (a) Meridional wind distribution; (b) streamlines of the total transport; (c) wind stress $\tau(y)$ against surface elevation in the meridional direction.

118

bringing the entire tropical circulation closer to reality.

More sophisticated models of the tropical current system were subsequently constructed, and these also incorporated inertia and the transition of the Coriolis force through zero at the Equator. These models will be discussed later. At any rate, the primary cause of equatorial countercurrents revealed by Shtockman – the meridional inhomogeneity of the wind – is predominant over all other effects.

6.4 Equatorial undercurrents

The discovery of the first equatorial undercurrent was a true sensation in oceanography. In 1951, a young oceanographer, Townsend Cromwell, headed an expedition to study the habitat of tuna fish in the equatorial Pacific. Special equipment was used for this study. Lines 16–19 km long were held on the surface by glass floats. Leaders with weights and hooks at the end were sunk from the lines to a depth of a few hundred meters. The line was sunk into the sea every morning and pulled out in the evening. The work was conducted in the area of the stable South Equatorial Current. The line was expected to drift with the water of the current to the west, But Cromwell noticed that it underwent a daily eastward displacement. Was this an error in taking the ship's bearings? Or was it perhaps the action of some random jet, just another antic of the sea? An ordinary mind would have been satisfied with this explanation, but Cromwell had the flair and curiosity of a real scientist. Immersing all the instruments that he had at hand to great depth, he discovered a mighty steady-state subsurface current. It was established later that he had discovered a current equal in scale to the Gulf Stream.

Cromwell was killed in an air crash in 1957 on the way to his new oceanographic expedition. The current he discovered has been named after him.

The discovery of a subsurface current in the tropical Atlantic was not so unexpected. In May 1959, a powerful easterly flow was found to exist under the thin layer of the South Equatorial Current. Between 1960 and 1967, researchers of the Marine Hydrophysical Institute of the Ukrainian Academy of Sciences, USSR, conducted detailed studies on this current and developed its theory. The current was given the name of the great Russian scientist, M. Lomonosov, after the name of the ship from which it was discovered.

It should be noted here that the Atlantic Equatorial Undercurrent was first noted by Buchanan during the 'Challenger' expedition in 1872–6. He inferred the existence of such a flow from temperature gradients beneath the Equator. Therefore the Soviet oceanographers actually rediscovered the Lomonosov Current.

In the Indian Ocean, the undercurrent was discovered in 1960 by Soviet research vessel 'Vytiaz' and later in 1962 was explored in detail by Americans on 'Argo', the research vessel of the Scripps Oceanographic Institution.

The existence of equatorial undercurrents in all three oceans indicates that it is a global phenomenon. We shall describe its main features. Undercurrents differ little in the various oceans. Everywhere they are directed along the Equator, crossing the ocean from west to east in a relatively narrow band symmetric with respect to the Equator (from 2°S to 2°N). The vertical thickness of the jet is 100–200 m. Current velocities show slight variations within a season. The jet coincides along its entire length with the upper thermocline, that is, the tropical density discontinuity layer. In the east, the layer filled by the undercurrent rises to the surface while its thickness diminishes. The greatest velocity of the undercurrent in the Pacific Ocean is 150 cm s^{-1}; in the Atlantic Ocean, 120 cm s^{-1}; and in the Indian Ocean, 80 cm s^{-1}.

An important feature of the Cromwell and Lomonosov Currents is their jet-like behavior. Their flow (34–40×10^6 m^3 s^{-1}) varies little along their entire course. This constancy is remarkable in view of the extremely large length of the jets, which in the Indian Ocean attains 4400 km; in the Atlantic Ocean, 5200 km; and in the Pacific Ocean, 33 000 km. Earlier we mentioned the jet-like nature of the Gulf Stream. However, in that current the volume transported varies quite drastically along the jet length. Whereas in the Strait of Florida the Gulf Stream carries 26×10^6 m^3 s^{-1}, its transport after coalescence with the Antilles Current is already 55–76×10^6 m^3 s^{-1}, and north of Iceland the North Atlantic Current is just 5×10^6 m^3 s^{-1}.

To understand the nature of the Cromwell and Lomonosov Undercurrents, it is important to determine the sources or the currents that supply water for their formation. This aspect has been investigated in some detail in the Atlantic. In carrying out his dynamic calculations of currents,

Defant long ago defined under the Brazil Current a flow directed along the South American coast and across the Equator to the north. It will be shown in the next chapter that this flow is a necessary element of the deep circulation in the Atlantic. Instrumental observations made from RV 'Mikhail Lomonosov' confirmed the existence of this flow at intermediate depths. Analysis of temperature and salinity fields suggested that this flow supplies water to the Lomonosov Current. Thus, the water of the undercurrent to a large extent arrives from the side.

A block diagram (Fig. 6.3) illustrates the 'incorporation' of the Cromwell Current in the system of the tropical zone. Without the undercurrent, the diagram is a modification of Defant's (1936) scheme, which appeared long before Cromwell's discovery. Included in this form in a great number of oceanographic texts and reviews, it creates an illusion of complete knowledge of the equatorial zone. Following works by Shtockman, Sverdrup and Reid, who explained the basic equatorial current system, theoreticians seem to have lost interest in tropical currents. Cromwell's discovery introduced new ideas which substantially changed the canonical scheme.

Following proof of the permanence of the Cromwell Current, dozens of studies appeared that discussed the effect of countercurrents from different standpoints. In order to systematize this variety of theories, scientists decided to compare the different opinions by presenting them all in one book. The editors of the international journal, *Deep Sea Research* (1960), invited well known oceanographers of that time – Stommel, Neumann, Charney, Robinson and Veronis – to present their analyses of the theoretical aspects of the problem. The fourth issue of the journal's sixth volume carries papers describing the greatly advanced notions of the mechanisms of the Cromwell and Lomonosov Currents.

These mechanisms are largely based on the 'shallow-water effect' described at the end of Chapter 4. It will be recalled that in a shallow sea the Coriolis force can be disregarded, as it is very small compared to friction. In the equatorial zone of the ocean, the horizontal component of the Coriolis force is also close to zero. Besides, the effect of the shallow 'bottom' is simulated by the density discontinuity layer, which is situated rather close to the surface at the Equator. The wind force, pressure gradient and friction are thus the only important factors that remain. The easterly

wind blowing over the surface along the Equator causes the drift current, which drives water toward the western coast of the ocean. As a result, there arises an eastward force of the pressure gradient due to the longitudinal slope of the free surface. It is this force that supports the undercurrent.†

These principles were clearly presented in Stommel's (1960) model. The model considers a narrow ocean band with edges symmetric with respect to the Equator. The currents are generated by a linear constant easterly wind. The surface (equatorial) current and the deep undercurrent in the plane of the Equator appear as in the diagram (Fig. 6.4). The only difference from a shallow sea consists in the fact that on Stommel's conventional bottom, which coincides with the tropical discontinuity layer, water particles do not adhere to the bottom but glide without friction. This scheme of motion is only observed in narrow bands near the Equator. At higher latitudes, the vectors of the upper current diverge under the influence of the Coriolis force (the parameter f increases linearly with latitude in either direction), and the deep countercurrent is concentrated near the Equator, as the velocity vectors converge here due to the Coriolis force. Although the real velocities in the countercurrent are almost twice as high as that calculated by Stommel, his model provides a good qualitative description of the phenomenon. The conditions of Stommel's model were quite rigid: the zonal total transport was supposed to be zero, that is, the amount of water carried to the coast by the surface current was equal to that removed by the undercurrent. This assumption holds well for uniform wind. If the wind is non-uniform, the Cromwell and Lomonosov Currents can be supplied from the side by waters of meridional currents. For instance, the Lomonosov Current, as we know, receives part of its water from aside.

J. Charney eliminated this shortcoming from Stommel's theory. His model includes the force of inertia as well as friction. The current velocity is assumed to be zero at the conventional bottom. In Charney's model, the resultant flow is not equal to zero. It is directed eastward. As a result, the deep

---

† These considerations are based on zero total transport in the zonal direction at vanishing Coriolis force. Logically the wind stress should produce a vertical circulation overturn, like that observed in a shallow sea (Fig. 4.11). Therefore, the Montgomery–Palmén model described above could not correctly explain the lateral turnover due to the same effect.

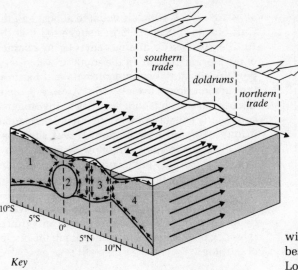

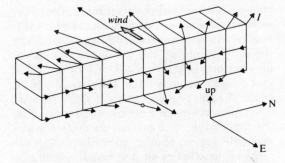

Figure 6.3 Winds and currents in the equatorial belt of the Pacific Ocean. Vertical scale is exaggerated substantially: the Cromwell Undercurrent should appear in section as a flattened lens.

Key

1 South Equatorial Current   3 Equatorial Countercurrent
2 Cromwell Undercurrent   4 North Equatorial Current

Figure 6.4 Schematic three-dimensional diagram of the flow field in the neighborhood of the Equator. The undercurrent is formed as a result of the surface pressure gradient (not shown). The flow is intensified just at the Equator due to convergence of water from both sides.

countercurrent is more powerful than in Stommel's model. Besides, part of the water can approach the undercurrent from the side. Despite this difference of approach, all theoretic models of Cromwell and Lomonosov Currents ascribe the dominant role to the longitudinal pressure gradient, which is offset by turbulent friction.

Important theoretic studies were done by Felsenbaum and co-workers (Felsenbaum 1970), who regarded the equatorial zone as a component of the overall oceanic circulation. Solving a succession of increasingly complex numeric models, they initially replaced a uniform ocean with a two-layer one, and then proceeded to an ocean with continuous stratification. Models with and

without inertia were analyzed individually. The best fit to observed data in the vicinity of the Lomonosov Current was achieved by a model of a stratified ocean with inertia terms playing a significant part.

There were several examples of additional narrow countercurrents at a depth of 200–300 m at the polarward limits of relatively homogeneous layers of equatorial subsurface water of about 14°C. In the eastern Pacific, these subsurface countercurrents are located symmetrically about the Equator, roughly at 5°N and 5°S, and are distinct from the surface countercurrents as well as from the Cromwell Undercurrent itself. Further to the west, the jets move closer to the Equator and have been observed to merge with the Cromwell Current. These flows are now regarded as permanent features of the equatorial current system there. Their existence is seen in the Atlantic Ocean as well. The nature of this flow is not yet clear.

6.5 Another unknown countercurrent

Undercurrents were not the only surprises to be found in the tropical zone. It was known that the North Equatorial Current in the Atlantic, on reaching the longitude of the Lesser Antilles, gives rise to the Antilles Current, while part of its water enters the Caribbean Sea to merge with the Guiana Current, which is a continuation of the South Equatorial Current (Fig. 6.5). The Guiana Current makes a turn for the north-west, going around the part of the South American continent that protrudes into the ocean. It was also known that north of the Equator all the currents in that area are directed westward.

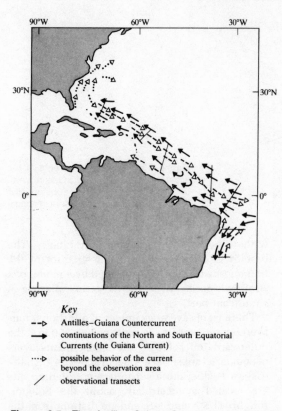

Key

--▷ Antilles–Guiana Countercurrent

⟶ continuations of the North and South Equatorial Currents (the Guiana Current)

····▷ possible behavior of the current beyond the observation area

╱ observational transects

Figure 6.5 The Antilles–Guiana Countercurrent (open arrows). (After Kort 1969.)

However, starting from 1961, eastward currents were measured north of Brazil and Guiana in a number of expeditions. An area with southeasterly currents was regularly found to exist on the transverse section of the southern part of the Gulf Stream, on its seaward side. In 1963 an Argentine expedition on 'Commodoro Augusto Lazerre' registered an easterly drift of the ship north of the Equator at 42°W. It remained unclear, however, whether that was a local anomaly or a new, unknown countercurrent. The problem was resolved after an expedition of the RV 'Academic Kurchatov', conducted in 1967 by the Institute of Oceanography of the USSR Academy of Sciences. All the cross sections from Florida to the Equator normal to the main coastline displayed a powerful countercurrent, which was given the name of the Antilles–Guiana Countercurrent by members of that expedition (Fig. 6.5). The countercurrent separates the long-known Antilles and Guiana Currents by a 130–240 km wide belt penetrating to a depth of 1000–1500 m. On the surface its speed is 50–100 cm s⁻¹. Volume transport in the middle portion of the countercurrent (at 50°W) is

30×10^6 m³ s⁻¹, which is equal to almost half the Gulf Stream transport. It was suggested that the Antilles–Guiana Countercurrent is fed by a branch of the Florida Current. In the south its waters support the equatorial countercurrents and partially the Lomonosov Current. The Antilles–Guiana Countercurrent has not yet been investigated sufficiently to determine whether it is a permanent feature of tropical circulation. It was suggested that the mechanism of this flow was based on a balance of Coriolis force and a pressure gradient transverse to the flow. It was not known, though, whether this gradient was a consequence of transverse wind non-uniformity in the region where trade winds meet the coastline or was only due to inhomogeneous thermal regimes along the circles of latitude.

6.6   Eddies in the subtropical Atlantic

As elsewhere in the ocean, the equatorial current systems possess eddies at various scales. They were first studied by Soviet oceanographers who employed the so-called 'Polygon' method. Basically it means that study of a dynamic phenomenon is concentrated in a characteristic area called a polygon rather than over the entire basin. The polygons are chosen as a result of a fast reconnaissance oceanographic survey, enabling the pinpointing of spots of maximum pressure gradients, regions of elevated vorticity, etc. Such spots are selected as sites to install networks of long-term recording current meters. This technique led to the discovery and description of the Lomonosov Current and the Antilles–Guiana Countercurrent.

In 1970, the USSR Academy of Sciences launched a large-scale experiment in the Atlantic Ocean code-named POLYGON-70 (mentioned in Section 2.5). A square with a side of 200 km chosen in the tropical zone was monitored for six months by mechanical letter-typing current meters attached at various depths to 17 moored vertical arrays positioned crosswise by the cardinal points. Taking part in the experiment were six modern research ships. Much care was taken in appropriate selection of the polygon sites. A relatively gentle bottom relief, sufficient distance from the coast and stable winds were among characteristics fitting the image of the open ocean. But the primary requirement was that the polygon should be within the area of the westerly, stable

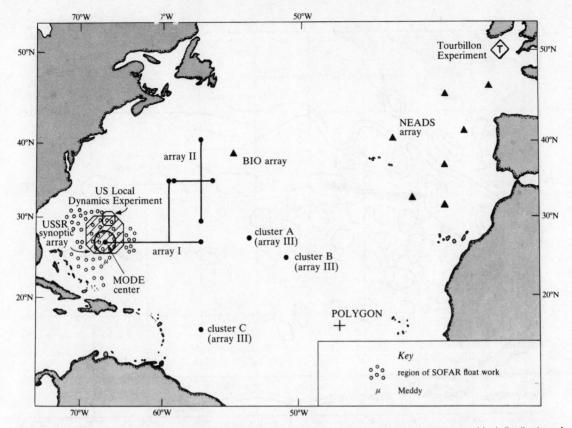

Figure 6.6 Location of recent major experiments on eddy currents in the North Atlantic – geographical distribution of POLYMODE and related fieldwork. (Courtesy of R. Heinmiller, US POLYMODE Executive Office, MIT.)

North Equatorial Current. One of the objectives of the study was to elucidate the relationship between 'constant' current and irregular ocean movements.

The results of the experiment were unexpected in many respects. First of all, no stable transport of water to the west was discovered. The directions of the currents at all levels varied sharply, many times, with a fluctuation period from one to six weeks. Only the average transport during the time of observation had a western component at most levels. The currents at two adjacent verticals separated by just 50 km showed considerable differences. The depth distribution of current vectors also lacked any obvious regularity. The fluctuating velocities substantially exceeded the mean velocity. Thus the duration of the experiment appeared to be not much larger than the timescale of a fluctuation. All this was observed in the zone of the trade winds, traditionally assumed to be the most stable area of atmospheric and oceanic motions.

Collected data showed that the polygon size was substantially smaller than the dynamic patterns passing through the network of measuring instruments. Only peripheries of huge circulatory patterns were monitored during the experiment. Assuming that these patterns are horizontal eddies, their diameters, according to estimates, should be about 200–400 km.

It has become clear that no observed average flow at the experimental site has enough energy to induce these immense transient phenomena, hence they have been attributed to some other external oscillatory source. It has been assumed that these are traces of the so-called Rossby waves.†

---

† The conceptual approach used throughout this book makes it almost impossible to discuss the nature of Rossby waves without the extensive introduction of the concepts of wave theory and related effects. Suffice it to say that Rossby waves are horizontal oscillations of properties induced by abrupt changes in wind systems. They slowly propagate in an ocean environment and their speed and behavior depend on the local rate of the Earth's rotation and stratification of the ocean.

123

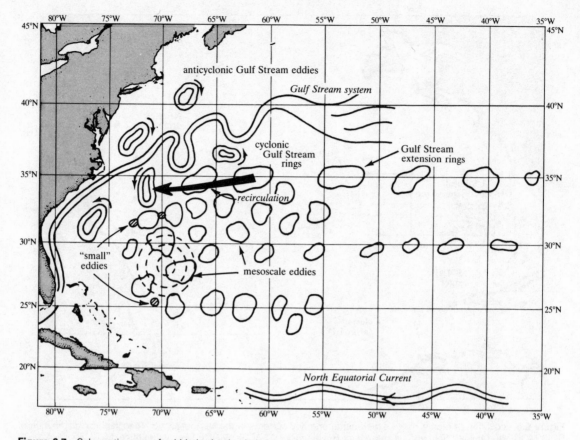

Figure 6.7 Schematic map of mid-latitude circulation and synoptic types of variable ocean currents – North Atlantic mesoscale variabilities.

We will not present here maps and diagrams of the first experiment of 1970, to save space for illustrations of the more detailed and larger-scale experiments that followed in the 1970s. In 1971–4, the international experiment MODE-I (Mid-Ocean Dynamics Experiment) was carried out in the vicinity of the Gulf Stream. The northwestern Atlantic Ocean became the scene of the most intensive scientific exploration. Figure 6.6 shows the sites of some of the various field projects: the Canadian (Bedford Institution of Oceanography) mooring array near the Gulf Stream, the US SOFAR float observational region during and after MODE-I, the NEADS (North-East Atlantic Dynamics Study) mooring sites maintained from 1976 to 1979 co-operatively by English, French and West German scientists, and the synoptic eddy study 'Tourbillon' (France–Great Britain in 1979).

It is hardly possible to encompass all the results that followed from the field experiments and numeric modeling. The reader is referred to the collection of well-written articles in the special issue of *Oceanus* (1976) and the recent treatise entitled *Eddies in marine science* edited by A. Robinson (1982).

The concept of Atlantic currents and eddies is schematized in Figure 6.7. The variety of eddies proved to be so great that their original causes and interaction with average currents appeared enigmatic. The most striking phenomenon is the behavior of relatively deep currents. Their study was made possible due to SOFAR floats (see Section 2.5) neutrally buoyant at 1500 m depth. The float paths (Fig. 6.8), which are called 'spaghetti diagrams', reveal a variety of complicated motions. At first sight, the motion appears chaotic, which corresponds to our concept of turbulence. However, a more detailed study shows that the floats reveal a number of circulatory mechanisms. This technique does not allow us accurately to isolate various processes, since the neutrally buoyant floats may change their vertical position during the course of the experiment. If ambient density is roughly determined by temperature, the

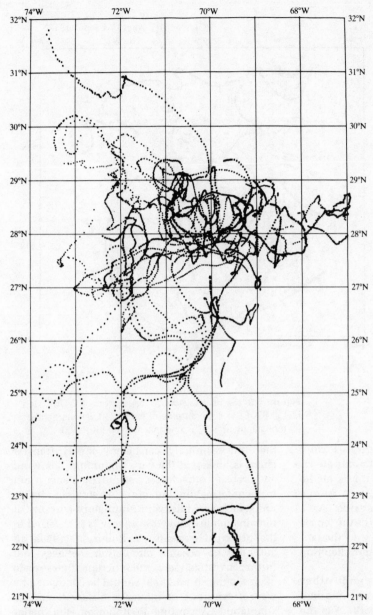

Figure 6.8 Spaghetti diagram of float paths.

possible vertical displacements of floats can be evaluated by maps of depths of isothermal surfaces. One such map is shown in Figure 6.9c. The small circles represent XBT stations. It is seen that the floats may dive 100 m over a distance of 200 km. Two maps of the MODE observational circle in Figure 6.9a and b demonstrate that the mid-thermocline eddy at 418 m depth does not extend far below. At 1500 m depth the pressure field is different.

Trajectories of floats sometimes demonstrate regular orderly behavior of water movement. This happened when floats were placed into a thin lens containing water with salinity and temperature different from the surrounding water mass. The high salinity of the lens left no doubt that this water came from the salty Mediterranean Sea. Therefore the striking anticyclonic eddy detected by floats within this lens (Fig. 6.10) was given the nickname 'Meddy'. The motion persisted for 8

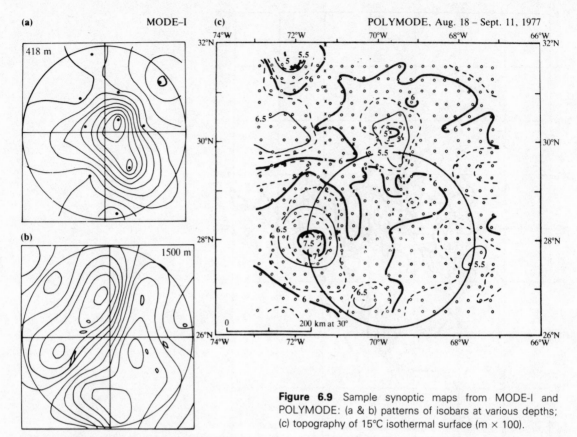

(a) MODE–I **(c)** POLYMODE, Aug. 18 – Sept. 11, 1977

418 m

(b) 1500 m

Figure 6.9 Sample synoptic maps from MODE-I and POLYMODE: (a & b) patterns of isobars at various depths; (c) topography of 15°C isothermal surface (m × 100).

months. The stability of the motion in this water mass seems to be related to the baroclinicity effect. Sharp temperature and salinity contrasts at the lens boundaries diminish the effect of energy transmission into the lens from outside. As a result, the movement of particles inside the lens is largely due to the pressure gradient – i.e. the volume force operating from outside and changing slowly.

As a result of POLYMODE and other experiments, oceanographers have obtained ample data to set their minds at work. Was the mechanism of formation of major oceanic currents unambiguous? So far, we have no doubt that the wind transmits its energy directly to the drift currents and that the piling-up of water creates a difference of levels between individual ocean regions, deforming the pressure gradient fields throughout the water body, to operate as the driving force in the deep oceanic layers. Proceeding from this scheme, we have explained all of the major oceanic currents, but it has been found

that the seemingly consistent energy transfer chain is broken at the very beginning. The wind and probably other forces transmit their energy not to major formations but rather to a series of smaller units such as eddies and undulatory currents. It remains unclear how this energy is transferred to the principal currents. A. Robinson (1980), an authority on oceanic circulation, believes that further investigation of the function of mesoscale effects in the current field should be conducted by way of synthesis of intensive observations and simulation. According to Robinson, the 'major longer range of objectives which should be achievable to a large extent by the end of the 1980s are: (i) the implementation of a Descriptive Predictive System for meso- (synoptic) scale currents over large regions of the ocean, which should consist of a numerical–dynamical model, a statistical model, and a real time observational network, including subsurface and remotely sensed components; and (ii) the construction and operation of a verified Large Scale Circulation Model with

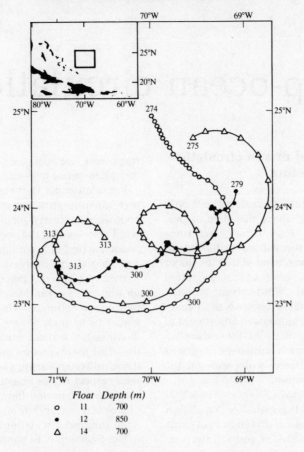

Float	Depth (m)
○	11	700
●	12	850
△	14	700

Figure 6.10 SOFAR float tracks in an intense, small mesoscale eddy.

effective and efficient parametrization of meso-(synoptic) scale processes for long time scale applications including coupling to an atmospheric model.'

7　Deep-ocean circulation

7.1　Effects of wind-driven circulation on ocean structure

Up to now we have discussed ocean currents only in the surface layer, i.e. above the main or permanent thermocline. This layer constitutes approximately 10% of the volume of the World Ocean. It was amply demonstrated that the major current systems in the upper strata are generated by atmospheric forces. Wind-driven currents penetrate into deep layers because of density gradients due to surface slopes and adjustment of temperature and salinity fields. All these are effective to a certain depth, which is different in various parts of the ocean and largely depends on the general density stratification.

Water movements exist at all depths. In Section 3.5 we briefly discussed thermohaline circulation in a deep meridional channel driven by heat gains and losses at various latitudes and salinity variations due to differences in evaporation and precipitation. Since the differences in the heat budgets between the hemispheres has been disregarded, the resultant circulatory patterns appeared to be perfectly symmetric around the Equator. It has been shown that multicellular circulations can cause a characteristic distribution of properties which is consistent with the real stratification patterns in the ocean. Thermohaline circulation in deep strata is a long-term response of the ocean to planet-wide differences in the heat budget and, therefore, largely determines the 'climate' of the World Ocean, whereas swift and changeable motions in the upper layers shape the 'ocean's weather'.

In the preceding chapters a simple superposition of wind-induced currents and thermohaline circulation has been tacitly adopted merely for convenience. It has allowed us to discuss separately the immediate effects of atmospheric forces upon the ocean, assuming that the 'bottom' of the surface layer involved in motion (i.e. the main thermocline) is formed by long-term processes of heating and cooling. Within this framework, we managed to explain major circulatory phenomena in the upper strata of the ocean.

If one attempts to investigate movements and thermohaline structure in the deep ocean, this approach is clearly insufficient. Obviously, the wind systems over the ocean are also climatologic factors, acting in accordance with the global heat and moisture budgets. The wind-induced movements have far-reaching cumulative effects on the entire ocean structure, thus affecting the deep-ocean climate. Therefore, the next step would be to analyze how the surface circulation modifies the vertical convective cells created by the ideal thermohaline process (Fig. 3.12). Since the wind systems are to a certain extent symmetric with respect to the Equator, we can, for a while, continue to ignore differences in the two hemispheres and concentrate on the effects within major wind-induced circulations.

The first step is to simplify wind systems over the ocean. We assume, as before, that the winds are primarily zonal. Recall that average climatic conditions (Fig. 1.5) suggest that the prevailing winds are from the east between the Equator and 30° (north or south) latitude. These are the trade winds. The region between 30° and 60° latitude is exposed to winds from the west (westerlies). Polarward of 60° latitude, weak easterlies should dominate.

Since the conditions in our ideal basin are assumed to be symmetric around the Equator, only the northern half of the ocean needs to be considered (Fig. 7.1). Now recall that convective overturn due to intense cooling in the polar latitudes is much more effective than convection due to salinity increase under the trades. In the latter case, intense heating in the Tropics prevents the density from being too high and limiting the depths of vertical descent, as opposed to the situation at higher latitudes, where the long-lasting winter cooling and release of salts during freezing act together to increase the density of surface water and hence the depth of convection. Therefore, we can, for a while, ignore salt convec-

(a)

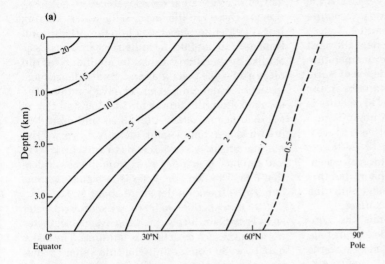

(b)

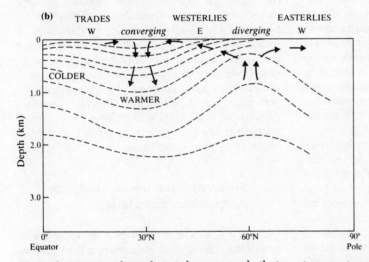

Figure 7.1 (a) Distribution of temperature in an idealized motionless ocean as a result of eddy conductivity. The temperature distribution at the surface is assumed to be a cosine function of latitude; the temperature along the vertical under the Pole is zero. (After Sverdrup (1940).) (b) Schematic representation of a meridional temperature distribution in an ocean (Northern Hemisphere) as a result of differences in the meridional surface heating, heat conduction and mass adjustment to zonal currents. E and W denote currents and winds flowing to the east or west, respectively. (Adopted from Neumann and Pierson 1966.)

tion in the Tropics altogether and assume only that unequal heating is responsible for the meridional temperature distribution. If further idealization is assumed and the horizontal translation is stopped, allowing only horizontal and vertical eddy conduction to redistribute the heat supply from the surface, we will arrive at the model developed by H. Sverdrup (1940). With the assumption that the surface temperature decreases from the Equator to the Poles and the temperature is nil along the vertical at the Pole, the resultant temperature field may take the form shown in Figure 7.1a. Though this model is completely artificial, it bears some resemblance to actual conditions in the ocean, namely increasing stratification in an equatorward direction.

If zonal winds also act upon such an ocean, the

temperature field becomes more complex. We can trace wind-induced effects qualitatively (Fig. 7.1b). As was demonstrated in Section 5.3 (Fig. 5.4), and then later in Section 6.2 (Fig. 6.1), water is piling up in the surface layer within the subtropical gyre, thus causing a line of convergence along the ridge depicted in Figure 5.4. Similarly, Ekman transport generates lines of divergence, between the easterlies and the westerlies and between the trades and the equatorial belt of low winds (doldrums). The vertical motions produced by converging or diverging waters adjust the density field to the currents. The simple thermal structure, as shown for a thermally driven ocean, will be largely distorted by zonal currents and the isotherms must assume a certain complex form in conjunction with wind-induced vertical

129

motions (Fig. 7.1b). The effects of vertical saline convection on the tropical meridional structure were discussed separately in Section 6.2.

The simple model above depicts one of the most important features of the general temperature structure in the ocean – the accumulation of warm water at depths down to 1000 m and more in the region of subtropical convergence. This feature is observed in all oceans. Another feature of Figure 7.1b, the divergence between easterlies and westerlies and the polarward sloping of isotherms north of 60°, is also observed in the real ocean, particularly in the Southern Hemisphere, but this effect is less pronounced, and the water under the divergence zone never reaches the surface.

The picture in Figure 7.1b omits the convergence in places where polarward currents (e.g. the Gulf Stream, Brazil Current, etc.) meet currents of higher latitudes (e.g. the Labrador Current, Falkland Current, etc.). It is not only that the simple 'collision' of two flows causes downward motion, but there is also another effect: mixing two waters of equal densities may cause an increase in the density of the mixture, facilitating a descending motion. This process will be explained later using the T–S diagrams. The geographic positions of various lines of divergence and convergence were shown schematically earlier in Figure 2.3.

All these considerations show that dynamic mechanisms associated with wind-driven currents are very important in the thermal structure of the ocean and in deep-ocean circulation. At least three vertical fluxes (namely, under subtropical convergence, tropical divergence and equatorial convergence) largely determine the local depth of the main thermocline, dividing the warm and cold spheres of the ocean. These motions in the warm sphere, together with salt convection in the Tropics, contribute to the well manifested sharpness of the thermocline at low and moderate latitudes.

Three other mechanisms drive water into and out of the cold deep layer. The most powerful source of deep water, i.e. polar convection, supplies water to the deepest part of the oceans. The subpolar fronts (the Arctic and Antarctic convergences in the Atlantic Ocean) drive vertical downwelling of water to intermediate depths, whereas subpolar divergence between these two sinks is partially responsible for the upward motion that brings cold water closer to the surface.

The description of the driving mechanisms in the cold sphere would be incomplete if two impor-

tant sources of rather dense water were not mentioned. These are the warm salty waters coming from the Mediterranean Sea into the Atlantic and from the Red Sea into the Indian Ocean. These seas act like evaporation basins, since minor river run off and unstable precipitation cannot balance huge losses of water from them via evaporation. The effluents from the Straits of Gibraltar and Bab el Mandeb deepen along the slopes and spread in intermediate depth and density layers. The Mediterranean water descends to 1000 or 1500 m and can be detected far away from the coast (Section 6.6). The Red Sea water is less conspicuous, but can be traced to depths of about 800 m.

With the exception of these two sources of water at intermediate levels, no deep water originates from the sea surface in the low latitudes. The major bulk of water coming into the cold sphere comes from the high latitudes. These motions are very slow. Since vigorous vertical mixing is inhibited by the thick permanent thermocline, the waters of various sources propagate in the cold sphere in the form of thin laminae, retaining their original properties over long distances. This strictly empirical fact had a special value for oceanographic research, because it allowed oceanographers to develop various analytical techniques to investigate deep-ocean circulation, the position and the origin of various water masses.

7.2   Some approaches to studying deep-ocean circulation

Historically, a coherent picture of the general pattern of deep flows was obtained from their properties mainly the distributions of temperature, salinity, dissolved oxygen and some other elements. The practical reason for not considering velocity measurements, as well, was the technical incapability of making them until recently. The properties are so stable in time that only slight differences have been noticed between deep-sea measurements taken many years apart. The data published in the 1960s by F. C. Fuglister (1960) look very much the same as those collected in the 1930s by G. Wust. This means that statistically significant information can be obtained from spot measurements of ocean properties taken in various years.

It is important to estimate which properties are better for the identification of water parcels of various origins. Comparing temperature and salinity,

(a) ISOTHERMS (°C)

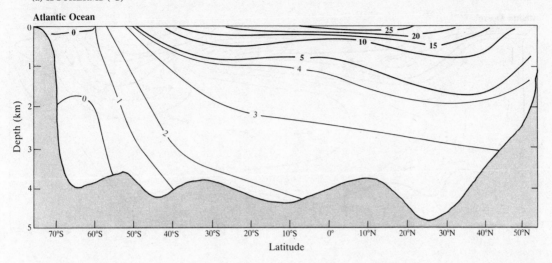

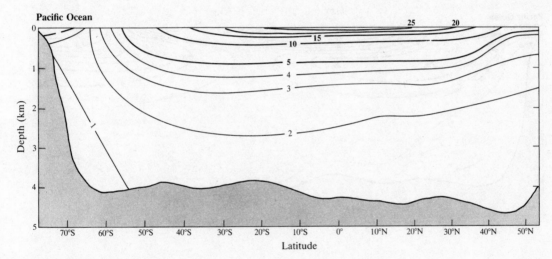

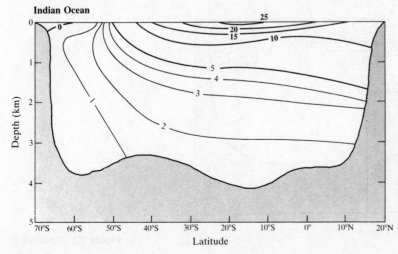

Figure 7.2 Vertical meridional distribution of (a) temperature and (b) salinity in the three oceans. Abbreviations for water masses are given in Table 7.1. Vertical arrows designate convergences and divergences: AAD, Antarctic divergence; AASC, Antarctic sub-polar convergence; STC, subtropical convergence; ASC, Arctic subpolar convergence. (Based on data collected by the Institute of Oceanology and published by Stepanov 1974.)

(b) ISOHALINES (‰)

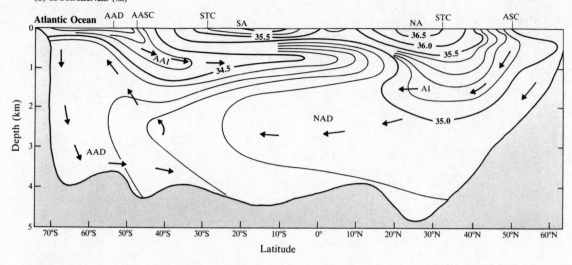

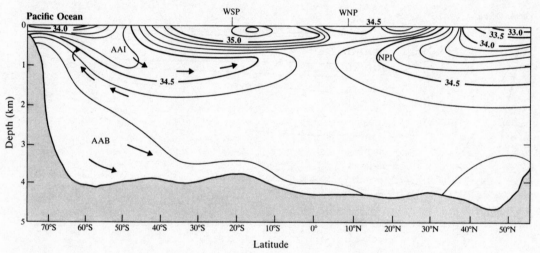

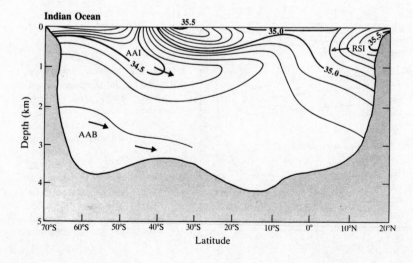

Figure 7.2 Continued.

one must admit that temperature is less suitable for describing the identity of a water parcel. Not only is this property less conservative than salinity but it also largely determines the density stratification of the ocean. Actually, for the range of temperatures occurring along the vertical (about 20–22°C in the Tropics), the changes in density are about 0.30–0.34% of its average value. The salinity changes from the surface to the bottom are only 1.0–1.2%$_{\circ}$, which contribute to density variations only 0.10–0.13% of the average density. Therefore, the laminated structure of the ocean is mostly controlled by temperature, and thus this property has limited use for identification of individual parcels of water. Salinity, in contrast, is a relatively good tracer. Not only is it a conservative property (there are no appreciable internal sources or sinks for salts in the water column), but also it is positively (due to evaporation) or negatively (due to precipitation) injected in definite zonal areas at the surface, where particular surface waters start their descent.

Oxygen is also a good tracer if consumption of it due to oxidation or respiration by marine organisms is known. Some man-made contaminants are even better tracers, for instance radioactive tritium, which has a vanishingly small natural concentration. Nuclear tests during the 1950s and early 1960s produced large amounts of tritium, and the temporal and geographic patterns of its input into the ocean have been well reconstructed. Elevated concentrations of tritium in some areas indicate relatively recent sinks of water. Another tracer of this kind is the radioactive carbon (^{14}C) naturally occurring in the sea. Beginning with the first fusion-bomb tests in 1954, the concentration of ^{14}C has increased sharply. Since that time the average amount of radioactive carbon in the atmosphere and upper ocean has been well documented. The amount of radioactive carbon in water depends on its age (the time elapsed since it left the surface) and the degree of mixing with ambient water. This ^{14}C allows one to determine the lengths of propagation and the speed of a body of water if the mixing rate is known or is negligibly small.

For a long time one of the basic indirect methods used to investigate ocean structure and deep-sea movements was the so-called 'core-layer' method developed by G. Wüst in 1935 (the name derives from the core of a flow, which is sandwiched between two layers of different properties, but the technique will require some clarification, and this is given later in the chapter). Wust identified seven such core layers, characterized by maxima or minima of dissolved oxygen, salinity, or temperature. Some of them can be distinguished in a very general cross section (Fig. 7.2b), where cores of low or high salinity are indicative of dominant flows. Though the path of a flow may not exactly coincide with the maximum (or minimum) value of the variable by which the flow is identified, this method provides information on the original characteristics of the body of water that forms the core. Comparing the salinity structures with temperature transects (Fig. 7.2a), one can see that temperature in general gives less information on water displacements than does salinity.

The stability of deep-ocean structure, so well manifested by its properties, was not confirmed by direct velocity measurements. Several current records obtained in the western part of the North Atlantic showed that a persistent current may change abruptly in speed and direction. Simultaneous measurements of salinity and temperature did not show that these changes in flow somehow affected the properties. It is not yet known how to reconcile the variability of currents with the steadiness of properties. Perhaps, some specific motions can occur on a much smaller scale within a nearly homogeneous body of water, not affecting the local water structure.

7.3 Classic scheme of deep-ocean circulation

The scheme of two large deep convective cells symmetric about the Equator with water sinking in high latitudes, flowing equatorward at depth and rising in the Tropics to return polarward near the surface was first published by the Russian physicist Lenz in 1845.† He formulated this principal mechanism of deep-ocean movements using the earlier work of A. Humboldt and very sparse data on subsurface vertical temperature gradients in the Atlantic and Pacific Oceans. Lenz even attempted to explain some specific traits of deep-ocean structure, such as shoaling of the ocean thermocline in the Tropics. He also recognized the role of wind-driven circulation in the substantial

† An excellent historical review on the important events and dates in the development of ideas about deep circulation has been published recently by B. Warren (1981).

distortion of deep motions from the simple cellular form.

When the data on deep observation of temperature had been amassed following the 'Challenger' expedition, Lenz's symmetric cells were eventually refuted. In the early 1920s A. Merz and later G. Wust proposed a comprehensive new picture of the meridional circulation in the cold sphere of the Atlantic. These works laid the foundation for the classic scheme of deep-ocean structure. The essential feature of this scheme was hemispheric exchange of water: an ocean-wide northward flow across the equator in the upper kilometer, compensated by a southward flow, with basin-wide northward and southward flows of bottom water in the western and eastern parts respectively.

Later programs for the Indian and Pacific Oceans, undertaken by L. Moller and G. Wust and then by H. Sverdrup, demonstrated that these two oceans were different from the Atlantic Ocean in terms of their deep-sea dynamics. In those two basins, hemispheric exchange is less significant. Though the exact mechanism of sinking was not known, the coldest and lowest salinity water at the foot of the Antarctic continental slope indicated that there should exist sources there of the bottom waters. These waters propagated northward. Above, a return flow was found, directed to the zone of Antarctic divergence. Further upward, this flow was overlaid by the layer of intermediate flow, again directed northward. This flow structure can be traced (Fig. 7.2) in the southern parts of all three oceans. However, the meridional flows in the Indian and Pacific Oceans appeared to be much weaker than those in the Atlantic and could hardly cross the Equator.

The differences in deep-ocean circulations of the southern halves of the oceans were correctly attributed to peculiarities of the coastal geometry and intensity of sinking in both hemispheres. The South Atlantic cuts deeply into the Antarctic continent in the form of the Weddell Sea, which is the coldest spot in the ocean. It was noted that bottom water comes from this sea and then spreads in the Antarctic circumpolar region to other basins. The sink under the Antarctic convergence supplying water to the intermediate layer appeared to be more powerful also in the Atlantic. This water together with the water from the deep source in the north were involved in the Antarctic Circumpolar Current, which provided a deep-water connection between the three oceans. Therefore, both deep

and intermediate waters entered the Indian and Pacific Oceans from the south, bringing measurably diluted waters from the Atlantic Ocean. The existence of several individual basins in both oceans, separated by well defined ridges, hindered northward propagation of water in the deepest layers. In spite of seemingly similar conditions in all three basins (nearly circular band of water, wide open expanse to the south, similar circulatory features, etc.), hemispheric exchange in the Atlantic Ocean appeared to be much more significant compared to that in the two other oceans.

In the north, the oceanographic conditions are completely different in all three basins. The Atlantic Ocean has relatively good connection with the Arctic basin through the Denmark Strait and the passage between Iceland and the British Isles. Though the exact paths of the downward motion were not known until recently, these passages were supposed to ensure close contact between the warm continuations of the Gulf Stream (the Irminger Current and Norwegian Current, Fig. 2.2) with cold water coming out of the Arctic basin. Apart from that in the cold Labrador Sea, the surface waters, saltier than in the other oceans, displayed a distinct sink as a result of convective overturn in the vicinity of Greenland or Baffin Bay. Arctic convergence, where the Gulf Stream continuations meet the cold and less salty water from the north, was also distinctly pronounced, and this was accounted for as a source of intermediate southward flow.

The Pacific Ocean is heavily partitioned from the Arctic first by the Aleutian Ridge and then by the shallow Bering Strait. Convective overturn and Arctic convergence were believed to be insufficient to provide a strong source of deep water. The very cold Sea of Okhotsk could produce some dense water in winter; however, this source has been proved by later observations to be insignificant as a deep sink. Therefore, the deep basins in the North Pacific were believed to be blocked from northern sources and, hence the entire body of water below the main thermocline was supposed to have an Antarctic origin.

The Indian Ocean is completely landlocked from the north, and the only source of water into the cold sphere might be Red Sea water. This can be identified by a tongue of elevated salinity (Fig. 7.2b). As in the Pacific, the bottom layers of the Indian Ocean contain waters once brought from the Antarctic continental shelf. Slightly elevated salinity in the southern parts indicates that the Atlantic deep water also participates in the formation of the Indian deep waters due to inter-basin exchange.

7.4 T–S relationships and water masses of the World Ocean

The description of water masses in the ocean became possible owing to the extensive use of T–S diagrams, briefly described in Section 2.1 (Fig. 2.1). This powerful analytical technique was introduced into physical oceanography by B. Helland-Hansen in 1916. Data on temperature and salinity obtained at any oceanographic station can be plotted on the T–S diagram, thus forming a T–S curve with depth ascribed to each point (Fig. 7.3).

Since a laminated water structure prevails over large geographic areas, and consists of only a few well-defined 'effluents', the shape of T–S curves does not change substantially from place to place. The surface layer, where salinity and temperature may change periodically, is not usually considered in analysis of the T–S curves. Below the depth of frictional influence, T–S curves preserve their form in typical regions so precisely that oceanographers routinely used T–S diagrams as a check on the tightness of their sampling bottles and the correct functioning of their thermometers. Any variations from the T–S plots typical for a particular area could be attributed to intrusions of alien water masses.

The concept of a 'water mass' is related to the T–S representation of ocean structure. According to Sverdrup, a water mass is defined by a segment of a T–S curve; in other words, a water mass is characterized by definite ranges of both temperature and salinity. Another term, a 'water type', is frequently used to designate a body of water with fixed temperature and salinity. According to definition, a water type is depicted by a point on a T–S diagram, whereas a water mass may consist of a mixture of two or more water types. Thus the T–S curves are a tool to analyse the mixing of two and more water types, and this also allows one to define T–S characteristics of a water mass at the place of origin.

If the vertical structure of the ocean consists of two water types of temperatures T_1 and T_2 and salinities S_1 and S_2, correspondingly, as shown in Figure 7.4, the T–S diagram reveals only two points. When these water types mix across the interface between the layers, the temperature and salinity profiles will undergo a certain smoothing, as shown by the broken lines. Then a water mass composed of a mixture of two water types will take the form of a line segment on the T–S diagram, because each variable in the mixture changes proportionally to the amount of water involved in mixing.

If three layered water types mix across two interfaces, the ultimate T–S plot will consist of two lines with a rounded angle at the water type 2 (Fig. 7.4b). This construction suggests that each observed T–S curve allows one to define the core of original water type. It will occur at the vertex of the T–S plot, as shown in Figure 7.4b. Following these cores in the open ocean, one can determine the spreading of water types and water masses and their relative stages of mixing. This is the basis of

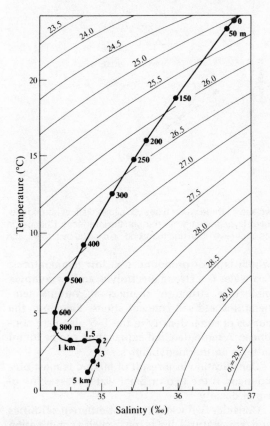

Figure 7.3 T–S diagram for average conditions in the center of the southern subtropical gyre in the Atlantic Ocean, between 20° and 30°S latitudes. (From data collected by the Institute of Oceanology, Moscow.)

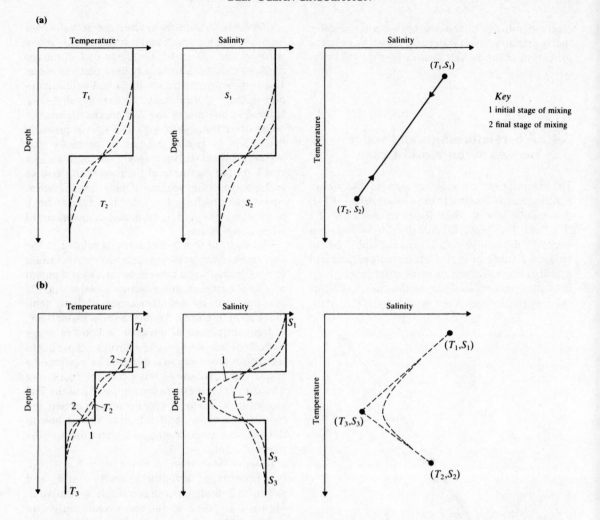

(a)

(b)

Key
1 initial stage of mixing
2 final stage of mixing

Figure 7.4 Schematic representation of vertical mixing of (a) two and (b) three water types. Originally distinct homogeneous layers of water with temperatures and salinities indicated in each layer gradually mix through the interfaces. Changes in vertical distributions are shown by lines. On the right the $T–S$ plots show evolution due to mixing: in case (a) by arrows, and in case (b) by two consecutive curves.

the 'core layer' method of Wüst, mentioned earlier. For instance, the $T–S$ characteristics of the core of water mass at a depth of about 800 m (Fig. 7.3) are $T = 3.7°C$ and $S = 34.35‰$.

Using the $T–S$ diagram we can explain why the density of the mixture of two water masses may increase.

This effect, sometimes called **densification**, is associated with some specific relations between the density of sea water and the temperature and salinity. At constant pressure, the density σ_t increases systematically with decrease in temperature and with increase of salinity, but the rates of these changes are not constant. The lines of equal σ_t (Fig. 7.5) are slightly concave upward,

which is most pronounced at low temperatures. Sea water is a strong electrolyte, and its complex molecular structure changes at various temperatures and salt concentrations. Therefore, the curves of equal density in the $T–S$ plane are non-linear. A more detailed explanation can be found in texts on thermodynamics.

Densification as a result of mixing is most pronounced if the two original water masses are of equal density.

Consider two water types of different salinities and temperatures, but taken in such a combination that the corresponding points on the $T–S$ diagram lie on the same isopycnal σ_t. One type designated 'L' has $T = 3°C$. $S = 33.51‰$, which is close to the

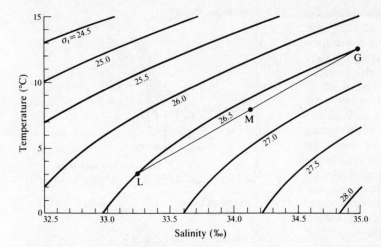

Figure 7.5 Effect of densification when two water types of equal density ($\sigma_t = 26.5$), L (Labrador Current) and G (Gulf Stream), are mixed in equal proportions. The mixture M lies on the straight line connecting these points and has density $\sigma_t = 26.65$.

conditions observed in February in the Labrador Current, when it approaches the northern edge of the Gulf Stream. The Gulf Stream water type at the same place is designed by 'G' and has characteristics $T = 12.6°C$, $S = 35.00‰$. These two types are represented by two points on the T–S diagram and are located on the same isopycnal $\sigma_t = 26.5$ (Fig. 7.5). When the two waters mix, the temperature and salinity of the mixture will lie somewhere on the straight line between them. If equal proportions of both water types are involved in mixing, the new water will have $T = 7.3°C$, $S = 34.2‰$. It is seen that the density of the mixture becomes slightly higher, as would happen for any water mass on the line LG. Thus, when two currents of waters of equal density converge and mix, the mixture will be denser and cause sinking.

The T–S method was extensively used in the 1930s to identify various water masses in the World Ocean. Excellent generalization was made by Sverdrup, Johnson and Fleming (1942) to describe layering of water masses that have common vertical salinity and temperature distributions, as denoted by their T–S curves (Fig. 7.6).

Sverdrup et al. distinguished two principal types of surface water masses: the central water masses between the subtropical convergences in the north and the south (Fig. 7.7) and the high-latitude surface water masses in the temperate zones between the subtropical and subarctic or subantarctic convergences. Since the high-latitude polar water masses form intermediate water masses, they are not shown on the map (Fig. 7.7). The arrows display the routes of sinking of the Antarctic and Arctic intermediate waters.

The major water masses have been divided into subtypes. Their names and the temperature–salinity characteristics typical of each ocean are shown in Table 7.1.

The central water masses are underlaid by the intermediate water mass, which originated in the region of the Arctic and Antarctic convergences. These water masses extend to a depth of 1500 m (Fig. 7.8). The Antarctic intermediate water is most widespread, and largely fills the major portion of all three oceans. In the northern parts of the other oceans, intermediate water masses are less conspicuous. The waters coming from the Mediterranean and Red Seas also fall into this category, but both waters at their oceanic origins are so homogeneous that they are regarded as water types. The tongues of the Mediterranean and Red Sea water types are clearly detected over large areas.

Further below, deep water masses are found. The most abundant pool of deep water is located in the Labrador and Irminger Seas. The North Atlantic bottom water propagates southward and rises over the more dense Antarctic bottom water (Fig. 7.8). The Antarctic deep water is identified in the Atlantic ocean (Table 7.1) because it is assumed to be climbing up between the two waters above and below. No deep water mass is formed in either the Pacific Ocean or the Indian Ocean.

The last and deepest waters are the bottom water masses. The only source of this water is Antarctic shelf water, which is spread in all three oceans. Later, we will find that there are very specific areas close to the Antarctic continent where the formation of the Antarctic bottom water mass actually occurs.

The T–S plots indicate the existence and characteristics of various water masses at each geographic location. But they do not by themselves describe the amount of water constituting each water mass. If this characteristic is also found, a three-dimensional T–S diagram can be constructed, where the height of each peak represents the volume of water with temperature† and salinity under this peak (Fig. 7.9). The idea of producing such volumetric diagrams was expressed by Montgomery in 1958, but its recent implementation was by Worthington. He collected data on temperature and salinity in the cold sphere and referred each station to areas of 5° of latitude and 5° of longitude. For each area, the abundance of water that exists in the ocean in each class is 0.1°C × 0.01‰, which was computed and simulated in the form of elevations in a three-dimensional sketch.

Only major water masses can be distinguished on the volumetric T–S diagram. Such well known water masses as the Antarctic intermediate water and the Mediterranean water are barely perceptible in the figure. But it demonstrates perfectly the differences between the bottom water masses in all three oceans. It is easily seen that the coldest and deepest water, which is formed in the Atlantic Ocean close to Antarctica, fills the vast bottom areas in the Indian and Pacific Oceans.

To summarize, in the classic framework the deep ocean structure was presented in the form of a layered construction, consisting of a limited number of primary building blocks – the water masses. The stability of the properties gave an illusion of rigidness to this construction, more similar to geologic strata than to the real motion in real water masses. The deep circulation of the cold sphere was considered to consist of ocean-wide flows in deep and bottom layers, though differences due to the existence of various deep ocean basins were recognized. The bottom water was considered to move generally northward and the deep water southward – with the greatest hemispheric exchange in the Atlantic and the least in the

† Note that potential temperature was used in this study, i.e. temperature from which a correction for pressure is subtracted. This is the temperature that a parcel of water would have if it is raised to the surface adiabatically, i.e. without heat exchange with the ambient water.

Figure 7.6 Temperature–salinity relations of the major water masses of the oceans. (From Sverdrup *et al*. 1942 by permission of Prentice-Hall, Inc.)

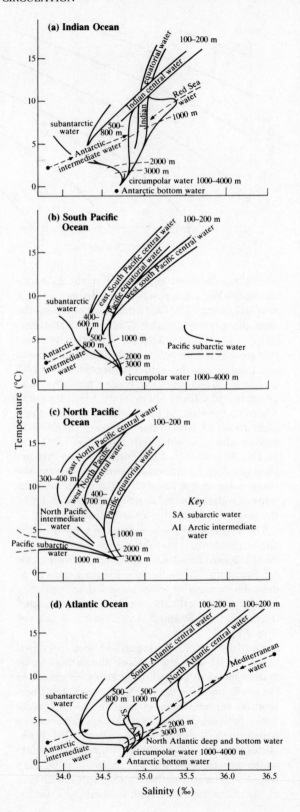

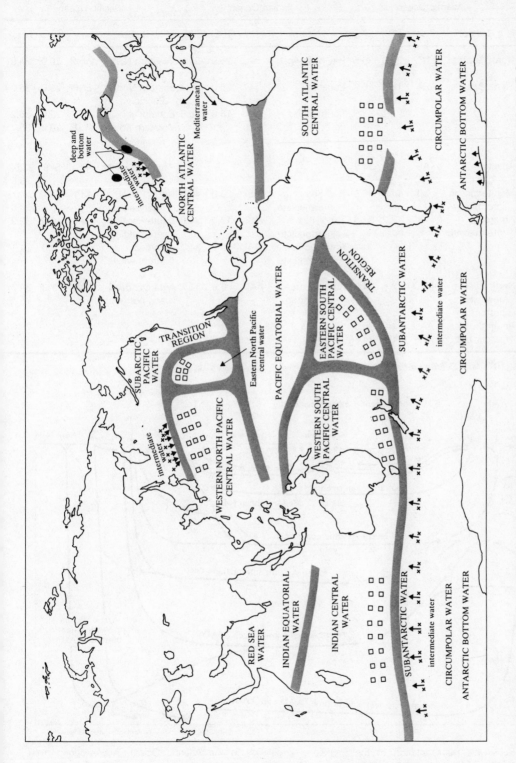

Figure 7.7 The geographic distribution of water masses that have common vertical salinity and temperature distribution, as denoted by their *T–S* curves (Fig. 7.6). Squares mark the regions in which the central water masses are formed; crosses indicate the lines along which the Antarctic and Arctic intermediate waters sink. (From Sverdrup *et al.* 1942 by permission of Prentice-Hall, Inc.)

Table 7.1 Major water masses of the World Ocean and their *T–S* characteristics.

Location	Atlantic Ocean Name	T(°C)	S(‰)	Indian Ocean Name		T(°C)	S(‰)	Pacific Ocean Name		T(°C)	S(‰)
central waters	North Atlantic	NA	20.0	36.5	Bay of Bengal	–	25.0	33.8	Western North Pacific	WNP	20.0	34.8
	South Atlantic	SA	18.0	35.9	Equatorial	–	25.0	35.3	Eastern North Pacific	ENP	20.0	35.2
					South Indian	I	16.0	35.6	Equatorial	–	25.0	36.2
									Western South Pacific	WSP	20.0	35.7
intermediate waters	Atlantic subarctic	AS	2.0	34.9	–				Pacific subarctic	PS	5–9	33.5–33.8
	Mediterranean intermediate	MI	11.9	36.5	Red Sea intermediate	RSI	23.0	40.0	North Pacific intermediate	NPI	4–10	34.0–34.5
	Antarctic intermediate	AAI	2.2	33.8	Timor Sea intermediate	TSI	12.0	34.6	South Pacific intermediate	SPI	9–12	33.9
					Antarctic intermediate	AAI	5.2	34.7	Antarctic intermediate	AAI	5.0	34.1
deep and bottom waters	North Atlantic deep and bottom	NAD	2.5	34.9	Antarctic deep and bottom	AAD	0.6	34.7	Antarctic deep and bottom	AAD	1.3	34.7
	Antarctic deep	AAD	4.0	35.0								
	Antarctic bottom	AAB	−0.4	34.66								

After Mamaev (1975) and Sverdrup *et al.* (1942).

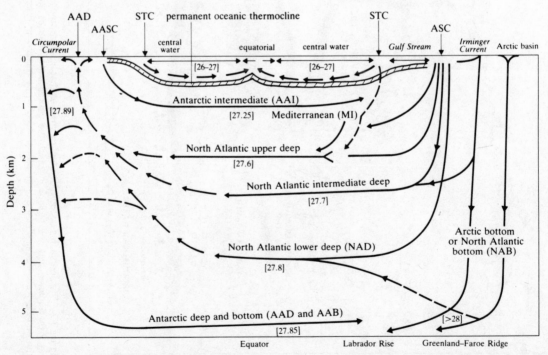

Figure 7.8 Schematic representation of the meridional circulation in the Atlantic Ocean. Abbreviations from Figure 7.2 and Table 7.1; σ_t is an average value.

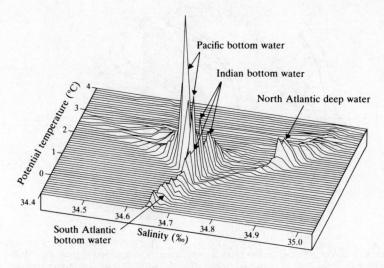

Figure 7.9 Computer-simulated T–S diagram of the water masses of the World Ocean. Apparent elevation is proportional to volume. Elevation of the highest peak corresponds to 26.0×10^6 km per bivariate class, $0.1°C \times 0.01‰$.

Pacific. The latest developments of this scheme demonstrated that only a few sources of deep and bottom waters exist in the ocean and these are highly localized. Sinking in the northern hemisphere was confined to high latitudes in the western basin of the Atlantic, with smaller fluxes from the Mediterranean and Red Seas. In the southern hemisphere, deep sinking was limited to the Antarctic continental slope in the Weddell Sea and, partially, in the Ross Sea.

7.5 Dynamic theory of abyssal circulation

The classic approach to deep-ocean circulation discussed so far was largely based upon empirical data and subsequent generalizations. In the 1950s a new theoretic approach was developed on the basis of the following two important concepts.

One of them concerns the mechanisms of water exchange through the upper boundary of the cold sphere. It was amply demonstrated that externally forced sinking is highly localized within small regions. A compensatory rising of cold deep water is obviously not as concentrated as the sinking, otherwise one should detect typical property distributions: for instance, well developed permanent divergence zones revealing upwelling of cold water to the surface. Since this is not observed in the open ocean (the coastal upwellings are shallow effects and should not contribute too much to deep circulation), we should assume that there exists widespread upward motion, which leads to mixing of cold water from below with warm water

from above. This entrainment of cold water into the warm sphere appears to be a necessary element of deep-ocean dynamics and this process carries away water from the top of the cold sphere, thus maintaining a water balance in the deep ocean.

The other concept also deals with common knowledge of ocean dynamics. Because of the very slow movements below the main thermocline, the turbulence is negligibly small in the cold sphere and the horizontal deep flow should be strictly geostrophic. This means (Section 3.4) that streamlines are parallel to isobars, and the speed of flow is inversely proportional to their spacing and the sine of altitude (see Eq. 3.7).

The concepts of concentrated sinks in the polar regions, uniformly distributed and compensating upwelling in the interior, and the geostrophic nature of deep horizontal flows were used by H. Stommel in 1955 to explain the basic mechanisms of circulatory patterns in the cold sphere. Later, Stommel and Arons (1960) constructed a mathematical model of deep global circulation, which we will largely follow.

The model consists of an ideal ocean of constant depth with homogeneous water rotating on the spherical Earth. The surface of this ocean represents the boundary (thermocline) between the cold and warm spheres. In order to be closer to the real world, the ocean is assumed to be bounded by two meridional walls from Pole to Pole, 60° of longitude apart. The motions are induced by two concentrated sources (S_1 and S_2) located close to the Poles (Fig. 7.10a) and the water is driven away from the sources by uniform upward motion of a known and constant discharge through the sur-

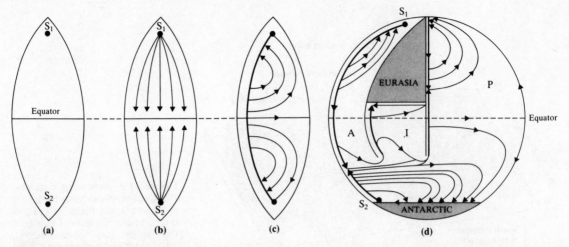

Figure 7.10 Theoretic scheme of abyssal circulation induced by the sinks S_1 and S_2 and homogeneous upwelling in the interior. (a) Positions of the source in the Pole-to-Pole ocean bounded by two meridional walls 60° apart. (b) Streamlines in the non-rotating ocean. (c) Rotation causes an intense boundary flow and a slow returning flow in the interior. (d) The deep circulation in an idealized World Ocean: A, Atlantic; P, Pacific; I, Indian Ocean.

face. This picture vaguely resembles the case of the Atlantic Ocean, with two sources of downward motion somewhere near Greenland and in the Weddell Sea.

If the ocean does not rotate (Fig. 7.10b), the streamlines of vertically integrated transport are not difficult to visualize. The downwelling waters will spread from their sources equatorward, being gradually 'pumped out' of the basin through the surface. A fan-like pattern of streamlines will be the most likely outcome.

When the ocean rotates, the circulation pattern loses its east–west symmetry (Fig. 7.10c). Since the motion is geostrophic, the equatorward flow will increase downstream, because its transport is inversely proportional to the sine of latitude. Some water should join this flow from aside, but it cannot grow continuously because the entire deep water of polar origin is eventually bound to rise in the interior and leave the cold sphere in each hemisphere. Therefore, the equatorward flows increase only over a segment of the flow path, where planetary effects, concentrating the current, dominate over the compensatory outflux into the interior. At a certain middle latitude, the southward flow begins to disperse due to the interior uptake.

Vorticity considerations similar to those used to explain the western boundary flows suggest that the intense equatorward flows should be narrow and located close to the western wall, whereas the interior flows are weak and diffuse. They are directed eastward and polarward.

Stommel and Arons translated their results to a multi-basin Earth approximately to three oceans connected in the south by an analog of the Antarctic Circumpolar Current (Fig. 7.10d). The oceans were represented as flat-bottom basins of equal depth bounded by either parallels or meridians. Again the sources were assumed to be only in the Atlantic Ocean, being represented by powerful deep-water sinks in the polar regions. The southern downwelling is also a source of water for the north-flowing western boundary currents in the Indian and Pacific Oceans, which enter them at their southwestern corners via the Antarctic Circumpolar Current. Therefore, the southern source supplies deep water to all three basins, instead of only one as shown in Figure 7.10c. The narrow boundary flow in the South Atlantic is less conspicuous and does not penetrate too far north, because the southern sink gives parts of their shares of the water to the Indian and Pacific Oceans.

Another phenomenon may contribute to inequality in the northern and southern abyssal circulations in the Atlantic basin. As indicated in Section 5.1, the intensity of the Gulf Stream system and the North Atlantic subtropical gyre is related to an excessive amount of water brought into the Northern Hemisphere by two (South and North) Equatorial Currents. There is no equivalent southward outflux of water in the east. The compensatory flow should take the form of local vertical sinks into the cold sphere, thus facilitating the deep western boundary flow in the Atlantic.

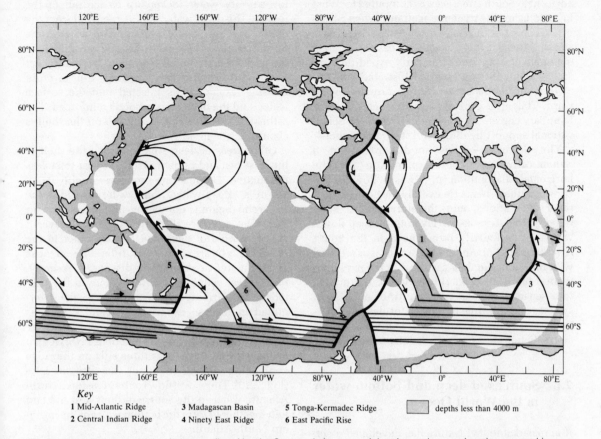

Key

1 Mid-Atlantic Ridge   3 Madagascan Basin   5 Tonga-Kermadec Ridge   depths less than 4000 m

2 Central Indian Ridge   4 Ninety East Ridge   6 East Pacific Rise

Figure 7.11 The abyssal circulation predicted by the Stommel–Arons model and superimposed on the geographic map. The locations of several mid-ocean ridges referred to later in the text are indicated. (After Stommel 1958.)

The resultant integral circulation pattern (Fig. 7.10d) shows the following major dynamic components: the downward motions feed directly (in the North Atlantic) or via the Antarctic Circumpolar Current the narrow equatorward flows at the western margins. These flows give rise to broad and diffuse circulations in the interior which lose much of their water in the compensatory, broad upwelling. A remarkable feature of this scheme are the interior flows that display recirculations in each hemisphere. This means that more water is involved in horizontal abyssal transport than is delivered by the assumed sinks. The resulting scheme of deep-ocean circulation transposed on to a geographic map is presented in Figure 7.11.

The model does not show any hemispheric exchange in the interior. Instead, the deep water in the open ocean of known polar origin should be moving on the whole toward its sources rather than away from them. This is a clear contradiction to the classic view of deep-ocean circulation,

which is based on basin-wide water exchange between hemispheres in the deep and intermediate layers.

The model predicts hemispheric exchange only within the narrow boundary flow at the western coasts. When H. Stommel (1955 and later) published the above conceptual scheme, with a deep narrow counterflow under the Gulf Stream as a basic element, the search was immediately begun to find this current off the US coast. Swallow and Worthington, in 1957, employed neutrally buoyant floats (Section 2.5) near the US Atlantic coast and found that the predicted current did exist. It appeared to be much slower than its surface counterpart. But the flow, with a speed of $1-18$ cm s^{-1}, was steadily moving south and south-west, opposed to the Gulf Stream above. This remarkable confirmation of the theory 'discovered at the tip of the pen' became one of the thrilling events of physical oceanography.

Incidentally, the relatively intense deep flow

adjacent to South America was computed by Wüst in 1935 using the dynamic method. When Stommel's ideas became known, Wüst in 1955 constructed two velocity sections based on data obtained during the 'Meteor' expedition in 1925–7. One of them vividly illustrated the Gulf Stream and its deep counterpart, flowing southward. Thus, the general idea of a deep western boundary current received comprehensive observational support by direct and indirect methods.

The Stommel–Arons model provided a dynamically consistent explanation for the horizontal distribution of abyssal flow fields. However, the scheme treats only vertically integrated circulations, and therefore misses such important features as the layering of deep flows, which is particularly pronounced in the South Atlantic (Fig. 7.8). Another drawback is associated with the flat ocean floor assumed in the model. In reality, the mid-ocean ridges divide the oceans into separate regions with multiple circulation systems and multiple entrances for deep and bottom water from the Antarctic, instead of single circulation systems in each hemisphere.

7.6 Sources of deep and bottom water in the World Ocean

It is remarkable that Nature has developed a certain 'bottle-neck' through which water is forced downward to induce deep-ocean circulation. The sources of downwelling are concentrated at very confined locations. Thus, oceanographers have received an amazing opportunity to investigate the rates at which sinking occurs using very limited technical means. In detecting the routes of deep water, one can also determine the layering of deep water and the amount of water admixed to the main thermocline. In spite of very inhospitable conditions in the Arctic and Antarctic, extensive efforts have been undertaken to investigate the areas of externally forced sinking.

Because of its predominant role, the Weddell Sea has attracted the most attention. A sink was identified east of the Bransfield Strait (Fig. 7.12) in an isolated trough filled with a mass of cold water different from the surrounding waters.

This newly formed water spreads in a plume-like manner along the bottom, and can be identified also by low salinity and low silica concentration. It was found that the prevailing offshore movement of the pack ice also transports low-salinity water locked up as ice out of the region. This light water is thus removed from the region of the sink and is mixed with water in the open ocean. At the same time, the movements of the pack ice leave wide areas open, facilitating net brine production at new openings due to rapid freezing. Some salt is injected into the surface water, and the deep convection is reinforced. The estimated total flux of dense water off the shelf is about 2×10^6 m^3 s^{-1}.

Other areas of deep sinking, along the shelf of the Ross Sea and Enderby Land, appear to be less productive. The first one produces roughly 0.6×10^6 m^3 s^{-1} deep water. Some water formed in the Antarctic region is not dense enough to form bottom water, so it spreads in the intermediate layers.

The sources of deep and bottom water in the North Atlantic are completely different from those of the Antarctic coast. As was discovered recently, the densest water does not originate in the Labrador and/or Irminger Seas, as it was earlier determined on the basis of very limited and sometimes unreliable measurements. The bottom water in the North Atlantic is formed of the dense Norwegian Sea water that flows over three sills on the ridge connecting Greenland and the British Isles (Fig. 7.13). These overflows entrain resident North Atlantic water in the course of their descent, and join together to form the bottom water of the northern North Atlantic.

The deepest passage of about 800 m is the Faroe Bank Channel, which allows roughly 1×10^6 m^3 s^{-1} water to flow along a quite complex route. After exiting from the Faroe Bank Channel, this current joins a second overflow, which passes over the ridge between the Faroe Islands and Iceland, where the sill is 300–400 m deep. This joint flow increases away from the ridge to 5×10^6 m^3 s^{-1} due to entrainment of the ambient water. On the way east, this flow meets the Mid-Atlantic Ridge, and passes through the Gibbs Fracture Zone (a deep ocean opening in the ridge), and then flows northward in the eastern Labrador Sea, and, south of Greenland, it joins the third overflow from the Norwegian Sea. That overflow comes through the Denmark Strait, where the sill is about 600 m. The Denmark Strait source provides about 5×10^6 m^3 s^{-1} water, which is distinctly colder and fresher than the Iceland–Scotland overflow.

The major routes of these flows are well documented and their strengths were found to be highly variable. Long-term current measurements demonstrated that overflow was occurring in

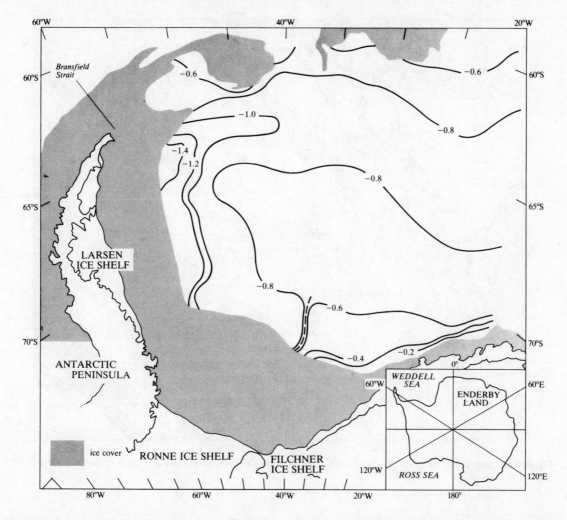

Figure 7.12 Distribution of bottom potential temperatures in the Weddell Sea, illustrating the eastward spreading to newly formed bottom water from the northern tip of the Antarctic Peninsula. *Inset:* the geographic names used in the text.

bursts of 1 or 2 days' duration. There are several reasons for these effects. Unlike the relatively steady convective sinking in the Antarctic shelf, the Earth's rotation induces local circulation in the Norwegian Sea, and spill from the sea occurs more as a result of general atmospheric conditions than due to direct cooling. Also, the heat losses and gains in this ice-free part of the ocean are highly variable during the year, and thus the deep flows themselves are variable.

The convective overturn in the Labrador Sea is less pronounced than that in the Norwegian Sea, but it has been noted several times. The depth of this convection is no more than 1500 m. Both the Labrador Sea water and the overflow water are more saline at their sources than is the Antarctic bottom water, and their salinities are increased on

the routes south due to mixing with the Mediterranean Sea water. The latter source produces about $1 \times 10^6 \ m^3 \ s^{-1}$ of warm but very salty water, injected directly into the North Atlantic deep water of the cold sphere. Because of relatively high temperatures the North Atlantic deep water overrides the Antarctic bottom water. The high-salinity Atlantic intermediate and bottom water is carried south into the circumpolar currents, and the elevated salinity is felt in the southern Indian and Pacific Oceans.

The Red Sea water enters through the Strait of Bab el Mandeb, having a discharge of $0.2 \times 10^6 \ m^3 \ s^{-1}$, and affects a relatively small area compared to that of the Atlantic, affected by Mediterranean outflow, adjacent to the Mediterranean Sea.

145

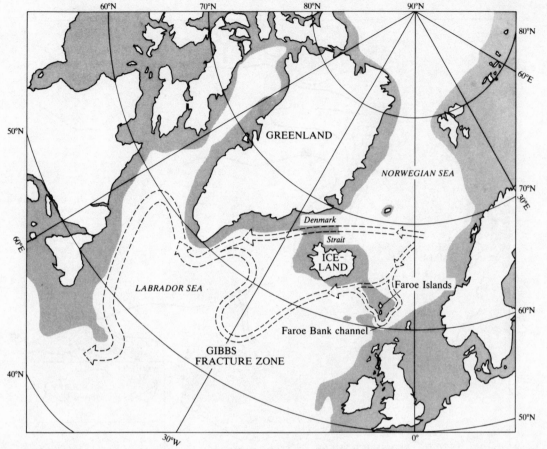

Figure 7.13 Index map identifying the overflow currents from the Norwegian Sea and local place names mentioned in the text.

7.7 Deep western boundary currents

We have outlined one of the important components predicted by the Stommel–Arons theory, i.e. deep localized pre-polar sinks largely controlling deep-ocean circulation. Two other elements of this scheme cannot yet be verified practically. It is clearly beyond modern technical capability to investigate the rate of slow vertical entrainment of cold water into the zone of the permanent thermocline. Some indirect estimates show that this rate is as low as 10^{-5} cm s^{-1}. It is technically feasible, though, to measure vertically averaged currents in the interior, but the appropriate program would be immensely expensive. However, the western boundary currents can be detected both using property fields (because of the geostrophic regime of the deep flows) or direct measurements.

The deep boundary flows are best studied in the Atlantic Ocean. A counterclockwise course of currents formed south of Greenland (Fig. 7.13) was determined by distinct slopes of isotherms and confirmed by float tracking and deep moorings in some instances. The width of this current was estimated to be about 200–300 m, in which 10×10^6 m^3 s^{-1} of water is flowing south-east.

Further south of Grand Banks, transverse temperature gradients do not mark the flow, but direct measurements by drifters and moored current meters clearly indicated that the flow does exist and is persistent. Near Cape Hatteras (lat. 35°N), its width is about 100 km, the volume transport ranging from 2 to 50×10^6 m^3 s^{-1}.

In the South Atlantic this flow is easily identified by its high-salinity core (more than 34.9⁰/₀₀; Fig. 7.14) between 1600 m and 2500 m.

146

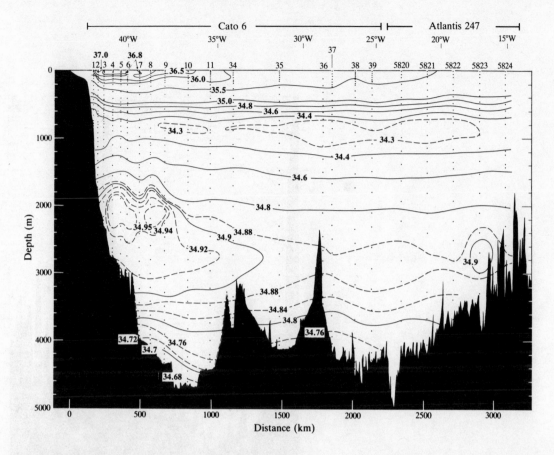

Figure 7.14 Section of salinity (⁰/₀₀) along roughly latitude 30°S from South America (*left*) to the Mid-Atlantic Ridge (see Fig. 7.10 for location), illustrating the two western boundary currents of the South Atlantic, namely the northward-flowing Antarctic bottom water and southward-flowing North Atlantic deep water above. Cato 6 (RV 'Melville') stations 1–11, November 8–12, 1972 and stations 34–39, November 25–9, 1972; RV 'Atlantic' stations 5820–5824, May 5–9, 1959.

The flow broadens to 500–100 km, perhaps due to lateral mixing effects. The volume transport decreases southward and constitutes about 9×10^6 m³ s⁻¹ between latitudes 8°S and 32°S. In the diagram (Fig. 7.14), a north-flowing boundary current can be identified by sloping isohalines 34.70–34.76⁰/₀₀. It carries approximately 6×10^6 m³ s⁻¹ of Antarctic bottom water and can be traced well beyond the Equator up to latitudes 8–16°N. It is worth noting that this current departs from the western continental slope and adheres to the Mid-Atlantic Ridge (Fig. 7.11). This transposition results, perhaps, from topographic effects, which may counteract the β-effect, as we discussed in Section 5.8. The northward bottom currents gradually taper off and near the Equator carry only 2×10^6 m³ s⁻¹.

The Mid-Atlantic Ridge has several noticeable fracture zones which allow some deep-water

exchange between the eastern and western basins divided by the ridge. No individual deep boundary flow has been found in the eastern basin.

The Indian Ocean has very complex bottom topography with a number of deep basins separated by ridges elongated in the north–south direction. Deep boundary flows were found along the eastern side of the Madagascan Basin (Fig. 7.15) and the Ninety East Ridge (for locations see Fig. 7.11). In both cases the flows are less well marked by transverse slopes of properties than are their counterparts in the Atlantic. Geostrophic calculations show that both flows have volume transports of about 4×10^6 m³ s⁻¹.

In the Pacific Ocean the western boundary current was found to be pressed close to the Tonga–Kermadec Ridge (for location see Fig. 7.11). The flow is marked by a faint salinity maximum below 3500 m (Fig. 7.16), which is the last trace of deep

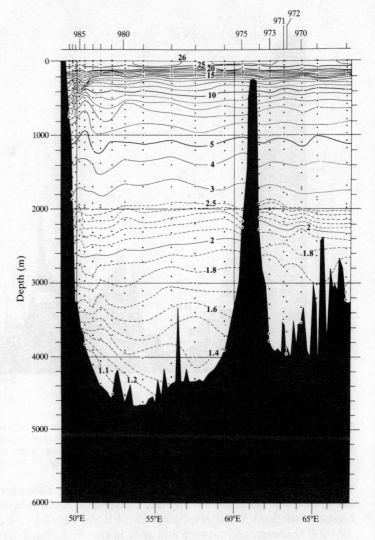

Figure 7.15 Temperature (°C) section along latitude 12°S between Madagascar (*left*) and the Central Indian Ridge, illustrating the deep western boundary current adjacent to Madagascar in the Mascarene Basin (see Fig. 7.10). RV 'Chain' stations 968–988, July 20–6, 1970.

water from the North Atlantic carried southward through the South Atlantic, and eastward around Antarctica in the Antarctic Circumpolar Current. The boundary flow in the Pacific carries about 20×10^6 m³ s⁻¹. It is probably the strongest of all the deep western boundary currents, and it is the major supplier of deep water to the largest ocean. Further north, this flow loses its identity, defined earlier by the salinity maximum, but other characteristics are indicative of the continuation of the northward propagation of the deep water.

The deep boundary currents near Japan cannot be identified by property fields, because the deep North Pacific is so far removed from near-surface sources. Some sparse velocity measurements indicate the existence of a recirculation region close to Japan, as predicted by the Stommel–Arons model (Fig. 7.11).

In summary, the existing observations clearly demonstrate that deep boundary currents are worldwide phenomena. The Stommel–Arons scheme provides a realistic description of the basic machinery controlling movements beneath the main thermocline.

148

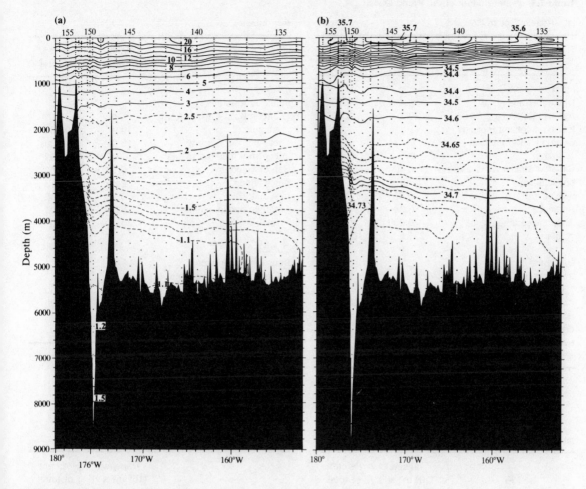

Figure 7.16 Sections of (a) temperature (°C) and (b) salinity (⁰/₀₀) along latitude 28°S between the Tonga–Kermadec Ridge (*left*) and the base of the East Pacific Rise, illustrating the principal deep western boundary current of the South Pacific. Scorpio stations 134–156, July 3–18, 1967, USNS 'Eltanin' cruise 29.

7.8 Water exchange in the World Ocean

The fluxes of water into the deep cold sphere are known very approximately. The time for complete renewal of cold water beneath the main thermocline (the **residence time**) can only be assessed roughly. The total volumes of the oceans (without inland and mediterranean seas) are shown in Table 7.2. It is assumed that the cold sphere constitutes 90% of the total volume in each ocean. The water is 'pumped' into the cold sphere by convective overturn in the pre-polar regions and by downwelling at subpolar convergences. For the entire ocean the total downward transport is approximately 34×10^6 m³ s⁻¹. These transports

Table 7.2 Volumes, transports and residence times of the cold sphere.

Ocean	Total volume (10⁶ km³)	Volume at the cold sphere (10⁶ km³)	Downward and lateral transport (10⁶ m³ s⁻¹)	Residence time (years)
Atlantic	338	304	21	460
Indian	292	262	10	830
Pacific	727	651	27	1030
World Ocean	1354	1246	34	1150

Table 7.3 Water budget of the World Ocean.

(a) Atlantic Ocean

	Transport (10⁶ m³ s⁻¹)	Percent of total
Influxes		
Drake Passage	183.4	79.20
Agulhas Current	30.3	13.10
Iceland–Scotland Passage	7.8	3.40
Denmark and Canadian Straits	6.1	2.60
Gibraltar Strait	0.70	0.30
precipitation	2.3	1.00
river runoff	0.52	0.22
Arctic ice	0.38	0.16
Antarctic continental ice	0.009	0.004
total	231.5	100.0
Outflows		
transport by Antarctic Circumpolar Current into Indian Ocean	214.9	92.80
discharge into Arctic	12.7	5.50
discharge into Mediterranean	0.89	0.40
evaporation	3.06	1.30
total	231.5	100.0

(b) Indian Ocean

	Transport (10⁶ m³ s⁻¹)	Percent of total
Influxes		
influx by Antarctic Circumpolar Current from Atlantic	214.9	91.50
East Australia Current	15.0	6.30
influx from Indonesian seas	2.06	0.92
Bab el Mandeb	0.16	0.06
precipitation	2.66	1.13
river runoff	0.19	0.08
Antarctic continental ice	0.017	0.01
total	234.9	100.0
Outflows		
transport by Antarctic Circumpolar Current into Pacific	201.2	85.50
Agulhas Current	30.2	12.80
evaporation	3.66	1.70
total	234.9	100.0

(c) Pacific Ocean

	Transport (10⁶ m³ s⁻¹)	Percent of total
Influxes		
influx by Antarctic Circumpolar Current from Indian Ocean	201.2	96.50
precipitation	6.55	3.29
river runoff	0.30	0.20
Antarctic continental ice	0.022	0.01
total	208.0	100.0
Outflows		
transport by Antarctic Circumpolar Current into Atlantic	183.4	88.20
East Australia Current	15.0	7.20
discharge into Indonesian seas	2.06	1.00
discharge through Bering Strait	1.14	0.55
evaporation	6.35	3.05
total	208.0	100.0

(d) Arctic Basin

	Transport (10⁶ m³ s⁻¹)	Percent of total
Influxes		
influx from Atlantic	12.68	89.40
influx from Pacific	1.14	8.80
precipitation	0.114	0.80
river runoff	0.149	1.00
total	14.01	100.0
Outflows		
discharge into Atlantic	13.8	98.20
transport of ice	0.190	1.40
evaporation	0.034	0.40
total	14.04	100.0

Computations were made by V. Kort, cited in Stepanov (1974).

are shown for various oceans in Table 7.2. Part of the Atlantic water is assumed to enter the Indian and Pacific Oceans.

The residence time averaged for the entire deep ocean may be calculated to approach 1200 years. A similar value was obtained from radiocarbon studies. The water exchange in the deep parts of the Atlantic Ocean is most vigorous due to two powerful sources; the Pacific Ocean keeps water in its deepest part much longer.

These computations are very rough and miss many local mechanisms that may facilitate renewal in the ocean abyss. Therefore, the residence times in Table 7.2 represent their upper limits.

Data on horizontal water exchange and land and atmospheric sources are much more reliable (Table 7.3). They indicate that the greatest water exchange occurs in the Antarctic circumpolar region, where 80–95% of the total transport is carried. Complete water renewal in these regions takes several months. Only a small portion of the circumpolar transport (about 7–15%) is involved in the interior circulation. Rough calculations suggest that the residence time in the upper layers within subtropical gyres may reach several years.

The Arctic Basin is renovated to the greatest depths (except for some deep basins far removed from the links with the Atlantic). The total volume of this basin is 17×10^6 km^3. The residence time of the Arctic due to water exchange indicated in Table 7.3 is about 38 years.

8 Currents and climate

At the beginning of this book, we artificially broke the cycle of continuous sea–air coupling. This allowed us to describe with sufficient approximation the large-scale mechanisms of atmospheric circulation. The ocean was involved in this analysis only indirectly as an inertial medium or buffer that must reduce the rate of global changes in the transport of heat, moisture and various substances. The representation of general atmospheric circulation in the form of a near-surface wind field has proved sufficient for the subsequent detailed analysis of oceanic circulation.

We now possess knowledge of the ocean that will enable us to analyze some mechanisms of interaction of the two fluid environments. We will be particularly concerned with both thermal and mechanical disturbances in the ocean which are strong enough to affect the atmosphere. Examining the ways by which these disturbances are transmitted through the entire system and the variations in the physical state of sea and air that they produce, we will arrive at an understanding of some natural mechanisms of climatic changes of the Earth and the effects of human activities at present and in the future.

8.1 Short-term cycles in the ocean–atmosphere system

It has been shown in Chapter 1 that vertical and horizontal movements of air are easily described by several heat engines set in motion by thermal contrasts due to non-uniform heating of air from below and fluxes of moisture and latent heat into the atmosphere. Let us now see how these processes operate together with motions in the ocean and are closed into common circuits in the sea–air system. The atmosphere, as well as the ocean, is penetrated by jets and eddies of different scales, from turbulent fluctuations (measures in seconds and meters) to cyclones and anticyclones (measures in weeks and thousands of kilometers), and even to huge 'stationary' patterns of general atmospheric circulation. This complex system moves, fluctuates and slowly rotates around the Earth.

Our everyday experience suggests that in some places weather can change overnight, whereas in other regions it might be rather stable in time. Some areas are frequented by hurricanes, tornadoes and floods, some areas are subjected to repeated droughts. Sometimes we become aware of longer variations of climate rather than just the weather. A case in point is the warming of the Arctic that was widely reported in the 1930s. In spite of inhospitable conditions in many places, various forms of life have been developed almost everywhere on the Earth, and no weather variations can completely destroy it.

In the relatively recent past, there were periods of climate when winter was much colder than today. Ship logs and monastery records show that a cold spell continued approximately from 1450 to 1850. This period is sometimes referred to as the Little Ice Age (some historians, for instance, believe that the failure of Napoleon's 1812 campaign in Russia was largely due to a particularly harsh winter in Eastern Europe at that time). According to Lamb (1969), during the Little Ice Age, pack ice approached Iceland, and the water temperature at Arctic latitudes decreased by 1–2°C compared to subsequent centuries. In the moderate and subtropical regions, in contrast, the temperature was higher. These variations, although noticeable, were not catastrophic.

Thus, the available data show that during historic time, that is, the past 4000 years, the state of the atmosphere and therefore of the ocean has been, on average approximately the same as at the present time. At least, annual fluctuations have never much exceeded the seasonal variations and temporary disturbances.

The geologic history of the Earth knows more profound changes of climate. During the past two million years, or the Quaternary period, considerable fluctuations of climate have occurred. Periods of glaciation (the glacial stages), when large portions of the planet were covered with ice, were succeeded by times of relatively mild climate (interglacial stages). Each such ice age continued from a few tens of thousands of years to hundreds of thousands of years. Ice sheets covered large parts of Europe, Canada and America, and pack ice extended to what are now temperate latitudes.

During the alternating glacial and interglacial episodes of the Quaternary, massive shifts occurred in the geographic distribution of plants and animals both on land and in the ocean.

Relatively recent glacial ages, particularly the last one, which reached its maximum some 18 000 years ago, affected the life of mankind. Human societies at low latitudes became largely dependent on domestication of plants and animals rather than on hunting, encouraging the gradual evolution of society in spite of the deadly ice covering some portions of the land, at higher latitudes.

In spite of drastic variations in the climate during the Earth's history, the average conditions so beneficial for life were preserved by many regulatory mechanisms in the sea–air interaction. Long-term variations in this system, leading to the ice ages, could be initiated either by cosmic factors, e.g. changes in the Earth's tilt, or for some terrestrial reasons, e.g. unusual volcanic activites. In any event, the ocean–atmosphere system, as we will see later, could develop some counteracting processes to offset the catastrophic trends.

On a smaller timescale, about hundreds of years, the cosmic factors (solar radiation and parameters of the Earth's orbit) remain fairly constant. For instance, during the past 50–70 years, no reliable variations in the solar constant have been registered (see Section 1.2) beyond the error of measurement. Any short-term instability in the ocean–atmosphere system is resolved within the system by several self-control mechanisms, which take the form of negative feedback effects. These phenomena should be discussed first.

Clouds function as one such control. When the sky is clear, the ocean is warmed more intensively, evaporation is enhanced, and clouds are formed. By screening individual areas of the ocean, clouds decrease the amount of solar insolation reaching the sea surface. Evaporation is diminished and, in the cooled atmosphere descending air flows develop in which the clouds condense. The process is then repeated (Fig. 8.1). These variations encompass largely planetary zones, including parts of continents.

B. L. Gavrilin and A. S. Monin developed a sim-

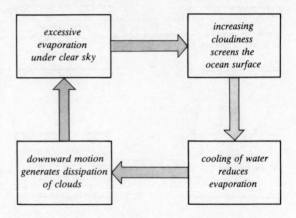

Figure 8.1 Self-sustained circuit with clouds as a flywheel. The approximate time of oscillations is a month.

plified mathematical model of heat transformations in a surface–cloud system and discovered a monthly cycle in their appearance (Monin 1972). This gives grounds to consider cloud systems as an important component of short-term weather anomalies.

Another powerful controlling agent are the ocean surface currents. Caused by wind, currents transport heat, eliminating or creating the temperature inhomogeneities in the atmosphere that give rise to wind. So far, efforts to develop even an approximate balance of the energy cycle of wind–currents–wind have failed. Probably, this cycle is too closely intertwined with other cycles.

P. Welander (1959) made an approximate calculation of the power resources of the subtropical gyre in the North Atlantic. He presented his results at the First International Oceanographic Congress in 1959. The work of wind stress in creating currents was estimated at 10^8 kW. At the same time, of the total amount of heat transported by surface currents, 10^{12} kW could be used potentially for non-uniform heating of the atmosphere. For production of wind, the atmosphere uses less than 1% of that amount, or 10^{10} kW, i.e. 100 times the amount required to maintain the system of currents. In addition to the 'broken' cycle, the different 'relaxation times' of the atmosphere and the ocean must play a role here. The atmosphere adjusts itself to variations of the ocean temperature field over several weeks, whereas the ocean sometimes takes several months to respond to average wind-field fluctuations. Perturbations in the wind–current–wind system prove stable and are quickly attenuated before completion of the entire cycle (Fig. 8.2). Apparently, currents must transfer

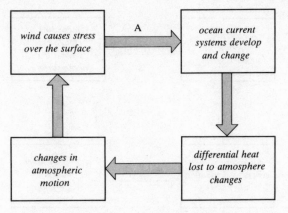

Figure 8.2 Possible stable feedback mechanism in sea–air coupling; oscillations die in link A because of inadequate frequency response of winds (weeks) and ocean current systems (months).

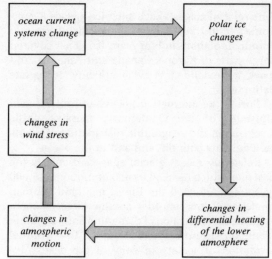

Figure 8.3 Schematic representation of the feedback circuit to 'store' the energy in changing the ice cover of polar seas. The system is self-oscillating because the ocean currents and polar ice have comparable frequency responses.

their energy to systems with the same degree of thermal inertia. Polar ice is one such system.

We will consider first the negative feedback mechanism in which polar ice plays a role as one of the controlling agents. This can be illustrated by an example. Imagine that for some reason the sub-tropical gyre in the North Atlantic becomes abnormally intensified. The North Atlantic Current will transfer more heat to the Arctic region, and its ice content will be decreased. As a result, the temperature difference in the meridional direction will be less pronounced, the pressure gradients decreased and the wind activity attenuated. However, when oceanic currents are also attenuated following the attentuation of the wind, so the process will reverse direction (Fig. 8.3). Shuleikin (1953) has calculated that in the North Atlantic the period of oscillations of average water temperature and ice content is approximately 3.5 years. Note that only the floating sea ice takes part in the process depicted in Figure 8.3, which is subject to seasonal fluctuations of thickness and density. The inertia of annually renewable sea ice can be compared to that of the top layer of the ocean involved in sea-sonal convection. The mobility of the sea ice is limited by friction at its bottom surface, by com-pression and by the rotation of the ice fields. An ever greater inertia is possessed by the perennial pack ice of the Arctic and Antarctic shelves due to its almost total immobility and great thickness (up to 4 m).

Huge ice sheets grounded on the Antarctic, Greenland and Canadian islands have a thermal inertia greater even than that of deep ocean waters.

In addition to three mobile spheres, the atmos-phere, hydrosphere and biosphere, and the immobile lithosphere, a separate semimobile entity is isolated called the **cryosphere**, which includes all the ice bodies and mountain glaciers. We will see later how the cryosphere is associated with long-range climatic variations. In the mean-time, let us discuss another negative feedback mechanism controlling the annual variations of the sea–air system.

In the mid-1960s, J. Bjerknes showed that, instead of the sea ice, the function of an inter-mediate link in the sea–air interaction could be performed by vertical water ascent in the ocean (Bjerknes, 1966). Bjerknes's paper, presented at the Second International Oceanographic Congress in Moscow in 1966, produced lively interest among oceanographers and meteorologists.

The reader knows that vertical transports may penetrate the entire water column in the ocean. At the depth of the thermocline layer they result from divergence or convergence of the Ekman layer, and deeper, from large-scale convection. It can be expected therefore that vertical circulation would regulate the heat fluxes of various scales. Bjerknes succeeded in analyzing the major consequence of vertical transports: the change of water stratification in the ocean, especially the intensity and depth of the ocean thermocline. For want of a constant monitoring service of the ocean, variations of the oceanic thermocline could not be observed with a sufficient degree of detail. Consequently, Bjerknes was prompted to his study by the behavior of a relatively small current, El Niño (see Fig. 2.2).

El Niño is a minor branch of the Equatorial Countercurrent passing along the coasts of Colombia and Ecuador toward the south. It meets the strong, cold, northwardly directed Peru Current. In normal years, north of the Equator El Niño turns into the open ocean and does not exercise any influence on the climate along the shores of Peru and Chile. Owing to the cold current there, almost no rain falls in a 100 km zone, and vegetation is extremely scanty. The sea, however, abounds in life. Waters rising from the depths carry nutrients. This is a region of strong upwelling.

There are certain years, however, when El Niño becomes intensified, penetrates far beyond the Equator and drives the Peru Current away. Strong winds storm the coasts. Heavy rains wash off soil and flood croplands and villages. In the littoral zone of the sea, the sharp temperature rise is fatal to cold-loving plankton and fish forms, and sometimes the population is stricken by epidemics. Disasters referred to as the El Niño events occurred in 1925, 1933, 1939 and 1944, and also in the winter of 1957–8. Later, in 1963, 1964 and 1965–6, the events were less catastrophic.†

For a long time the causes of the El Niño events remained unclear. Bjerknes correlated sundry data on water temperature with calculations of geostrophic currents and average annual maps of atmospheric pressure during the 1957–8 event and the three preceding years. He found that the particular conditions of El Niño were created during an attenuation of the easterly wind system east of the tropical zone in the Pacific. This led to a decrease in the rise of deep water and an overheating of the surface waters near the Equator. The Cromwell Undercurrent rose to the surface at the eastern coast of the ocean, intensifying the El Niño Current. The anomalous heating of the atmosphere in the Tropics aggravated the cyclone in the northeastern Pacific and led to intensification of winds in temperate latitudes.

Remarkably, the Atlantic during that period

† The most recent El Niño event in 1982–3 had, perhaps, one of the worst adverse effects on the global weather, causing severe droughts in Australia, Central America and other regions, and devastating floods in USA and Equador. Estimated damage reached $8.65 billion. (*National Geographic*, February 1984, 144–83).

experienced an attenuation of westward transport in moderate latitudes and a decrease of the zone of action of the Icelandic low-pressure area, with the entire Arctic falling within the scope of action of the high pressure over Alaska. As a result, the eastern part of the equatorial zone in the Pacific during the El Niño event became the 'atmospheric kitchen' for the entire Northern Hemisphere. The normal situation is usually restored after the excess of warm water at the Equator is removed by transport in the oceanic circulation. Temperature contrasts move closer to the Equator, and the trade winds are again intensified, leading, in turn, to a higher current velocity in the equatorial zone. The entire process of operation of the self-sustaining mechanism with the vertical rise can be illustrated by a diagram (Fig. 8.4).

Subsequently (for example, see Namias 1976) the processes in the air and sea preceding the El Niño episodes and their inverses (abnormally cold Peru Current) were studied in great detail. The findings confirmed Bjerknes's theory. Averaged sea level pressure distribution preceding the El Niño episodes (Fig. 8.5) shows a deep and southwardly displaced Aleutian low-pressure area. Above-normal pressure tends to be observed at low latitudes, which is a manifestation of a weakening of the trade winds and the subsequent events leading to the El Niño warm advance on the Peru coast. This chain of events gives a clue for predicting the El Niño episodes by studying the

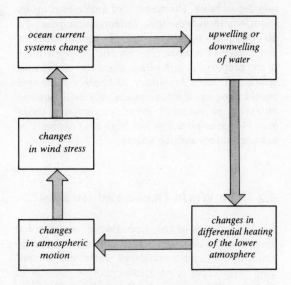

Figure 8.4 Three-dimensional feedback mechanism involving deep ocean water for differential heating of the atmosphere.

155

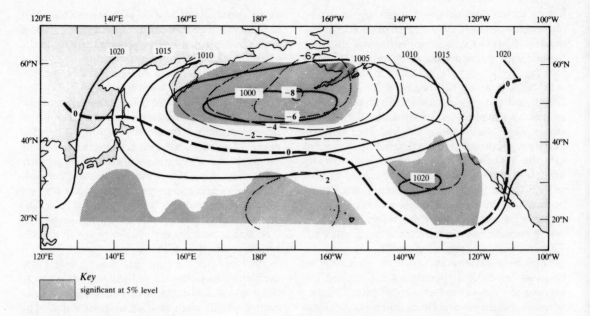

Figure 8.5 Average sea level isobars (full curves) in the North Pacific for the seven warmest summers (January–March) in the Southern Hemisphere and differences (broken curves) between those pressures and the mean for the seven coldest periods (in millibars).

changes in atmospheric circulation of the eastern Pacific as a whole.

These are relatively fast cycles of self-sustaining negative feedback mechanisms. Although it is not always easy, in many instances they can be revealed with sufficient reliability. Present-day views on longer processes are more difficult and less dependable. The scarcity of data leaves much room for alternate theories. Indeed, during the past 50 or 60 years a great number of original ideas on the causes of long-range climatic changes have been proposed, including theories of glacial epochs. A mere inventory of these hypotheses would take up a lot of space. We will confine ourselves to some of them, particularly those based to some extent on the high heat capacity of oceanic waters and ice sheets.

8.2 The World Ocean and ice ages

All the theories of the formation of a stable ice sheet over the Earth are based on the specific properties of the transformation of water into ice and vice versa. This is an almost unique process in Nature, with a positive feedback effect within the ocean–atmosphere system, which means that any initial thermal disturbance is amplified within the

system to a long-range change. For example, a snowfall or formation of ice greatly changes the reflective capacity of the Earth's surface, known as its 'albedo'.

The sea ice is known to reflect up to 80% of incident solar radiation, while the dark oceanic water absorbs solar heat very well. The albedo of the sea surface varies from 6 to 30% depending on a great number of factors, such as the angle of incidence of the solar ray, the roughness of the sea surface, the turbidity of the water, etc. Other insulating properties of the ice also lead to increasing temperature contrasts. The thermal exchange between sea and air is interrupted almost completely, and the air is cooled even more.

Another circumstance should be taken into account in the analysis of the thermal barrier during disappearance of the ice. When the ice melts in summer, the fresh water that is formed takes a long time to mix vertically. So, in spite of decreasing albedo, accumulation of heat in the region free of ice occurs only in the surface layer. On land, where the heat capacity of the underlying surface is negligibly small compared to that of the sea, the ice sheet is very stable. Not only floating ice, but also the entire cryosphere, takes part in increasing the ice sheet. The ice is a constant condenser of water, decreasing evaporation many times over. In dynamic terms, huge ice masses, located symmetrically relative to the Pole, lead to the formation of stationary polar anticyclones, which even more weaken the meridional circulation. The thermal isolation of the ice sheet is increased even more .

We have listed the factors responsible for the cooling of the Earth's surface following the formation of a stable ice sheet. In the Northern and Southern Hemispheres these processes, of course, take different forms due to sharp distinctions in the distribution of water and land around the Poles. To understand the physical mechanisms leading to the Earth's glaciation, however, these particulars can be disregarded.

One initial impetus, such as prolonged cooling, is thus sufficient to trigger a glaciation epoch like those which occurred in the Quaternary. Assuming that the influx of solar radiation in the upper envelope of the Earth is constant, we can suggest two possible causes of global glaciation. The first factor is variable amount of radiation received by different latitudinal xones of the Earth during individual seasons. This may result from the changeable position of the Earth's surface relative to the Sun, which depends on the eccentricity of the Earth's orbit and the inclination of the Earth's axis of rotation with respect to the plane of the orbit at the equinox. All these astronomic factors could contribute to changes leading to variations in the amount of solar energy received by different latitudes. These changes can be calculated adequately for the past millennium.

Another possible reason may be variation of atmospheric transparency due to changing amount of dust, mainly because of volcanic activity. The contribution of atmospheric transparency to absorption of solar energy can be crudely estimated from geologic data on volcanic activity during past epochs, but there is no evidence for a significant correlation between vulcanicity and global climatic variations of a magnitude to create or terminate the global glacial events. However, the possibility remains that spectacular volcanic dust clouds (Sandorini, Krakatoa, Mount St Helens) may have caused perturbations in the wind fields, which may have led to smaller perturbations in the surface current systems.

Each of these factors can change the radiation balance of the Earth to a limited degree. Because of this, a great number of authors doubted their potential influence. In the past few years, pre-

ference has been given to astronomic factors, particularly to orbital variations as a primary cause of ice age climatic cycles.

The process of onset of glaciation may be studied by using numeric models. One of the first such models was produced by Milankovitch (1941). An ice age, according to Milankovitch, begins when summer insolation decreases to such an extent that large portions of snow and ice survive from the preceding cold season. The subsequent spreading of ice becomes imminent due to the positive feedback effects as soon as the global albedo increases. The initial impact of reduced solar heating is associated with the precession of the slightly eccentric Earth axis with its tilt being low at this critical summer season. (For a detailed explanation, see Ruddiman and McIntyre (1981).)

The atmospheric and oceanic circulations were neglected in this model, though the growth of ice sheets over the land and the following ablation of ice cannot be explained without these processes. In recent models more importance has been assigned to the ocean.

The Russian climatologist Budyko (1972) developed an empiric model of a temperature regime of the glacial epoch, based on heat balance in the atmosphere and proceeding from a known global albedo of the Earth's surface, α. The main equation links the known solar radiation at the atmospheric boundary, Q, with the long-wave back-radiation into space, I, and the total heat flux, C, due to horizontal motions in the atmosphere and hydrosphere. In its most elementary form, this equation appears as

$$Q(1 - \alpha) - I = C \qquad (8.1)$$

The values of I and C are expressed by empiric functions in terms of the temperature of the Earth's surface, T, which is assumed as unknown in the model. Another empiric equation links the unknown boundary of polar ice with the planetary albedo. Data on the initial variation of solar radiation were taken from the above-mentioned work of Milankovitch.

Budyko's model disregarded the variation of radiation properties of the ocean and fluctuations of the amounts of clouds in the past as compared with the existing regime. However, his calculations indicated that during the last glaciation 18 000 years ago the ice sheet went south on average 8° along the meridian in the Northern Hemisphere and northward by 5° in the Southern

Figure 8.6 Sea surface temperature, ice extent, ice elevation and continental albedo for Northern Hemisphere summer (August) 18 000 years ago. Contour intervals are 1°C for isotherms and 500 m for ice elevation. Continental outlines represent a sea level lowering of 85 m. In northern Siberia, dotted lines indicate a recently revised estimate of the ice extent used in this experiment. To aid in the visualization of the thermal gradient along the North Atlantic polar front, alternate contour lines have been omitted in the western Atlantic. Albedo values are given by the key.

Key

loess, steppes and semi-deserts, albedo 25–29%

forested and thickly vegetated land, albedo <20% (mostly 15–18%)

snow and ice, albedo >40% – isolines show elevation of ice sheet above sea level

sandy deserts, patchy snow and snow-covered dense coniferous forests, albedo 30–39%

savannas and dry grasslands, albedo 20–24%

ice-free ocean and lakes, albedo <10% – isolines show sea surface temperature (°C)

Hemisphere. The temperature at the 'critical' latitude of 65° fell 5.2°C during the warm season. These results were consistent with the concepts current at that time as to the natural conditions during glacial epochs, obtained by geologic research.

A shortcoming of the model is its stationary character. A series of consecutive approximations was used to estimate the temperature and the ice boundary corresponding to stable external factors. Fluctuations in the system due to transfer of heat and moisture through the sea and ice surface were accounted for only in an indirect fashion.

Over the past decade or so, major advances in paleoclimatology have been achieved due to the development of quantitative methods for estimating past climatic variables. The multidisciplinary program known as the CLIMAP project (CLIMAP, 1976) was carried out as part of the International Decade of Ocean Exploration, funded by the US National Science Foundation. A consortium of scientists from many institutions participated in this program. The main goal of the endeavor was to reconstruct the paleogeography and paleoclimatology of the Northern Hemisphere at the peak of the last glaciation about 18 000 years ago.

This study also has practical applications. Human activities such as the burning of fossil fuels and overgrazing of semi-arid regions may trigger climatic changes at a rate completely different from those in the past. The theories of future climatic changes can be best tested by comparing them with the changes recognized in paleoclimates.

The most fascinating aspect of this grandiose project, however, is neither the three-dimensional models of the ice age climate (Gates 1976) nor the time-dependent models of planetary glaciation for the past 500 000 years (Imbrie and Imbrie 1980), but the fine analyses of microscopic fossils of ancient plankton retrieved from the cores of geologic samples. These methods, called the 'biological transfer functions' and 'oxygen isotope stratigraphy', allowed the compilations of global charts of sea-surface temperature and volume of paleo-ice. Following our principle of dealing only with thermodynamic processes, we refer the inquisitive reader to other sources (Drake et al. 1978 or CLIMAP 1976) for details of the stratigraphic technique.

Figure 8.6 shows the sea-surface temperature and maximum ice cover for the late Quaternary Earth 18 000 years ago. The most striking feature

of the world at that time was the Northern Hemisphere's ice system, consisting of land-based glaciers, marine-based ice sheets and either permanent pack ice or shelf ice. This ice mass stretched across North America, the polar seas and parts of northern Eurasia. In contrast, large Arctic areas in Alaska and Siberia remained unglaciated. In the Southern Hemisphere the winter extent of sea ice was significantly greater than it is today.

In the North Atlantic, the Gulf Stream shifted slightly southward off the Carolina coast and swept across the basin to Spain. The noticeable crowding of isotherms at 42°N indicates the southern edge of the polar water mass. There was no strong northward transport of warm, high-salinity waters such as occurs at present in the North Atlantic, due to the Norwegian and the Irminger Currents (Fig. 2.2).

The temperature changes are better examined on the anomaly map (Fig. 8.7). Maximum temperature anomalies (−18°C) are found approximately at 40°N in the west. They diminish toward the eastern coast and are pronounced near Spain. Generally, the central parts of the glacial subtropical gyres, though displaced toward the low latitudes, experienced only slight temperature changes. The only exception was the North Pacific (Fig. 8.6), which perhaps resulted from the shift of the Kuroshio, which took a more southern track in glacial times.

The ancient Benguela Current was also different from its present form. It is indicated by lower temperature plumes (Fig. 8.6) along the eastern coast of Africa and South America. Well developed negative anomalies (−6°C) in the low-latitude Pacific Ocean apparently resulted from increased upwelling along the equatorial divergence (Fig. 2.3). The CLIMAP members argued that the shallow equatorial thermocline (Fig. 6.1) perhaps broke down during the glacial interval, resulting in upwelling of cool waters to the surface. Recalling the Bjerknes sea–air mechanism (Fig. 8.4), we can understand what a tremendous impact these irregularities had on climate. Numeric simulation of climate at the peak of glaciation (Gates 1976) demonstrated that air temperature decreased much more drastically than did ocean temperature.

Processes of ice disappearance during the interglacial period have been studied only recently. In Figure 8.8, the major feedback mechanisms by which the ocean enhances the rate of ice disintegration during warming periods are

159

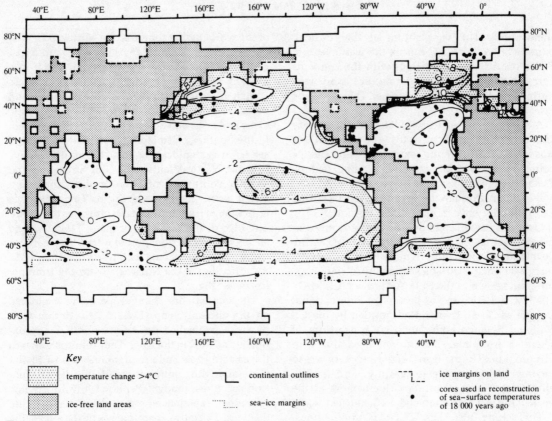

Figure 8.7 Difference between August sea-surface temperatures 18 000 years ago and modern values. Contour interval is 2°C. Continental and ice outlines conform to a grid spacing of 4° latitude by 5° longitude.

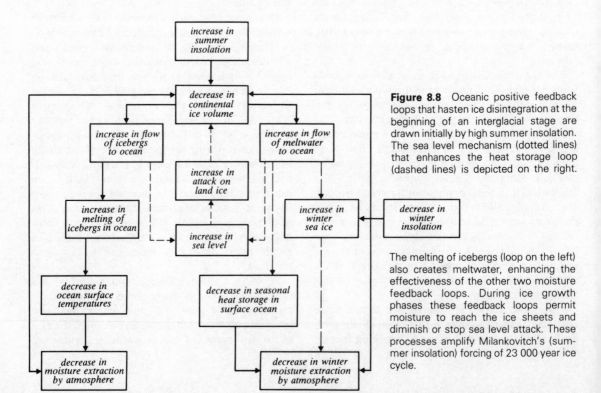

Figure 8.8 Oceanic positive feedback loops that hasten ice disintegration at the beginning of an interglacial stage are drawn initially by high summer insolation. The sea level mechanism (dotted lines) that enhances the heat storage loop (dashed lines) is depicted on the right.

The melting of icebergs (loop on the left) also creates meltwater, enhancing the effectiveness of the other two moisture feedback loops. During ice growth phases these feedback loops permit moisture to reach the ice sheets and diminish or stop sea level attack. These processes amplify Milankovitch's (summer insolation) forcing of 23 000 year ice cycle.

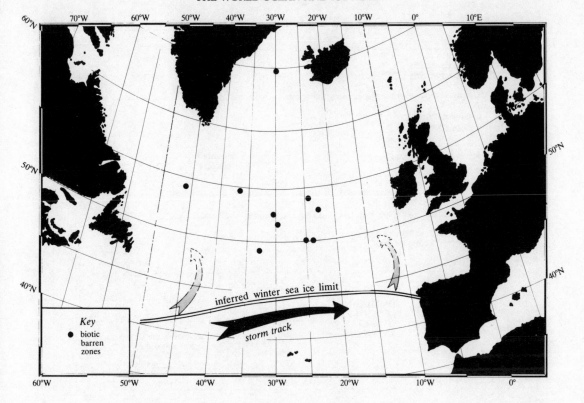

Ice-decay sequence

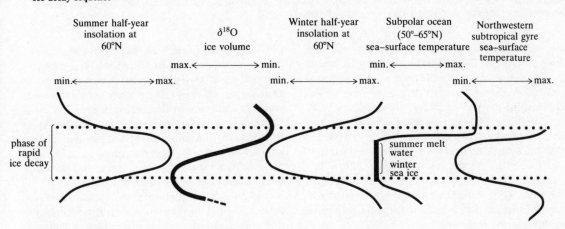

Figure 8.9   Map shows location of cores (circles) with biotic barren zones indicative of summer meltwater layer and winter sea ice during deglaciation. Arrow indicates winter storm track eastward pack ice limit with possible diversions into the ice-covered area.

summarized. The important loop at the left involves iceberg melting with subsequent changes in the vertical stratification and moisture supply to the ice sheets. At the same time, strongly stratified waters enhance winter ice formation and decrease the heat capacity of the pre-polar ocean, which

further amplifies the reduction of moisture flux (the right loop).

Note the mechanisms of thermal regulation in areas covered with sea ice which are new to us. Transport of icebergs, their gradual melting and the formation of a thin, less saline water layer

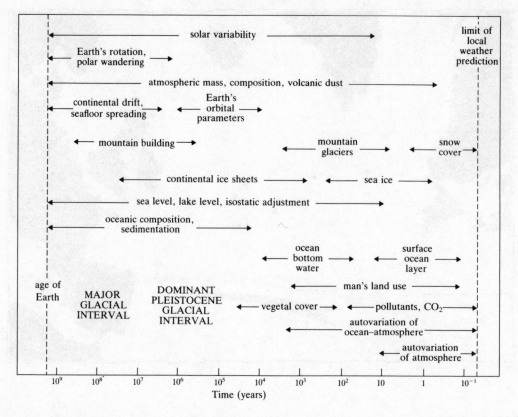

Figure 8.10 Characteristic climatic events and processes in the atmosphere, hydrosphere, cryosphere, lithosphere and biosphere and possible causative factors of global climatic change.

should confine the storage of summer heat to a shallow layer of the mid-latitude North Atlantic Ocean. At the same time the cold sea temperature reduces moisture flux to the land ice sheet, diminishing its growth. The increased sea level facilitates this loop by hastening iceberg calving. This is how the ocean amplifies the positive feedback once the deglaciation process has started. It was found that storm tracks tend to follow the strongest albedo boundaries and thermal gradients (Fig. 8.9). Therefore, when a storm moves polarward it is cut off from its source of energy and is weakened.

The impact of paleo-oceanic circulation on the process of deglaciation is not known. It is hypothesized that the intensified subtropical gyre might be displaced to the south. The prehistoric Gulf Stream had to act like the modern one to push the sea ice further northward.

A spectral analysis of paleoclimatic temperatures and ice volume has indicated that a 23 000 year cycle of ice growth and decay may prevail in the history of the Earth's climate, though longer cycles were also revealed. The primary cause of the 23 000 year cycle – the attenuation of summer radiation – is likely to occur in our time as well. The climate of the Earth, therefore, can be vulnerable to variations of thermal balance. Apart from the astronomic reasons described above, another powerful factor is exerting an increasingly more active influence here: industrial development, discussed in greater detail later.

In conclusion of this section, it appears useful to make a logical correlation and comparison of the probable duration periods of the factors that conceivably influence climatic change (Fig. 8.10). Remarkably, fluctuations in the sea–air system are of the order of 10^3 years. If the ocean bottom water is involved in this process, the time period of variation may reach 10^4 years. Climatic change due to feedback effects in the air–sea–ice sheet system attains on average 10^5–10^6 years, but it is not ruled out that the period of this variation can be as long as 10^8 years (100 million years).

8.3 The ocean and the problem of global power production

Assessment of uncontrolled climatic change† due to human industrial activity, especially the increased output of energy, has become particularly important today.

People have always experienced an energy shortage. In this age of industrial revolution, hunger for energy is more acute than before. This is evidenced, for example, by the fact that during the past 30 years mankind has spent half of all the energy utilized throughout history. It has been estimated that power production amounts to 0.016 W m⁻² energy (Bolin 1977). Comparing this figure with the difference between absorbed radiation and long-wave radiation (for the Earth it is on average 240 W m⁻²), one might think that it is too early to speak of effects of energy production on climate. Yet in regions with dense population and highly developed industry (tens of thousands of square kilometers), artificial production of energy attains 10 W m⁻² (for example, the North Rhine–Westphalia industrial complex), and in urban areas (tens of square kilometers) up to hundreds of watts per square meter. Manhattan, New York, releases 630 W m⁻², which exceeds six times the mean solar radiation at the Earth's surface. The entire energy of photosynthesis accumulated through geologic time (coal, gas and oil) is released, affecting substantially the climate of urban areas and larger districts.

The increase of energy output amounts to 4–5% annually (Budyko 1972). According to other estimates (Bach *et al.* 1980), the rate can be actually higher (up to 6%). This means that not later than in 100–200 years' time, the amount of heat produced by man will be commensurate with the amount of radiation balance of the entire surface of the continents. The air temperature will begin to rise globally. But this will not be the only effect of energy production. Heat from fire remains the main source of energy. When mineral fuel is burned, carbon dioxide is released, which results in the greenhouse effect described in Chapter 1. Besides, the burning of coal and wood leads to the release of a certain amount of dust and particles

† The book with the eloquent title *Inadvertent climate modification*, ed. K. L. Wilson (1971), contains facts and presents the results of studies conducted by scientists in various countries to estimate the effect of industrial activity on various layers of the atmosphere and water resources of land and sea.

which will screen solar radiation. Let us consider first of all how the global temperature will be changed as a result of energy production.

Using his model of the atmospheric heat balance, Budyko (1972) has calculated that with the present-day rates of growth of energy production, the temperature of the planet will begin to rise rapidly in the middle of the 21st century. By the year 2050 it may grow by 1–2°C, which will affect the climate greatly. The polar ice in the Northern Hemisphere will retreat, by the year 2000, by 2° latitude northward on average, and subsequently, during the next 80 years, the Arctic sea ice may disappear altogether. A gradual destruction of the ice sheet of Greenland and Antarctica will set in, leading to a rise of the World Ocean level. As a result, the freshwater balance on the continents will be strongly modified, leading to catastrophic consequences for the economies of many nations.

However, Bolin (1977) believed that energy production a hundred times larger than that of today would be required to change the global temperature by about 1°C. Nevertheless, secondary effects, such as increased evaporation and release of carbon dioxide, appear to be even more important. Keeling (1980) attempted to quantify the effects of fossil fuel carbon dioxide on the world's climate. Though future fuel consumption is uncertain, any increase (even less than 1.5% per year) will lead eventually to a large increase in atmospheric CO₂. If a major fraction of minable coal is used as fuel, a four-fold increase in CO₂ concentration is likely to be reached within three centuries. Keeling demonstrated that the ocean will be a less effective sink for CO₂, at the time of maximum CO₂ release, than it is now. The carbon dioxide cycle involving the ocean and land biota will fail to respond quickly to the increased amount of CO₂. This lag will lead to a sharp air temperature rise.

It is possible that the increasing amount of dust in the atmosphere due to energy production could also have global consequences. With increasing air pollution control, this effect can be diminished. It is believed that an increased particle content may slightly reduce the warming of the atmosphere caused by the greenhouse effect due to atmospheric carbon dioxide. To date there are no models that allow calculation of the contribution of various processes to the modification of climate by human activity. However, proceeding from the feedback processes described above, an approximate idea of the ocean's response to the new conditions can be developed. Let us consider some of the possible changes.

Warming of continents in the Northern Hemisphere will affect the operation of the heat engine of the second kind (see Ch. 1). In winter the contrasts between the ocean and the land will diminish and meridional circulation will become less intense. Stationary baric patterns in winter will be less contrasting. They will embrace larger territories since the land during winter will not differ from the ocean in its thermal characteristics as sharply as it does today. Yet in summer the contrasts must be intensified. The land warmed by energy production will still receive more heat from the sun, whereas the ocean due to its inertia will retain the winter cold. As a result of enhanced contrasts, a more intensive atmospheric circulation will be created.

The subsequent course of energy transformation will depend on the ocean's response. It can be assumed that in winter during attenuated winds the water exchange between the northern and southern parts of the oceans will be intensified in the Northern Hemisphere since meridional contrasts will be retained to a greater extent. In summer, on the contrary, wind-driven currents will be intensified and circulation on the surface ocean layers will be restored and probably increased. To what extent the modification of horizontal transport will affect the variations of the thermocline is difficult to predict. Yet, the rise time and disappearance of the El Niño episodes gives grounds to believe that the thermocline will be able to produce a timely response to seasonal fluctuations of the wind field. This means that part of the excessive heat created by energy production will not only melt the ice of polar regions but also, to some degree, warm up the ocean water mass and increase evaporation. Enhanced transport of salinized water to the north may lead to increased convection in the zone of subpolar convergence. Therefore, the deep layers of the ocean beneath the thermocline will begin to be warmed. The excessive atmospheric heat will thus be absorbed by the ocean and the time of melting of the Greenland and Antarctic ice sheets will be postponed.

Will the future climate be more humid than today? It can be assumed that in winter, with small thermal differences between the sea and the air, but substantial meridional temperature gradients, the transport of air masses from the ocean to the land will be increased, leading to a greater humidity of air over the land. In summer, sharp contrasts at the ocean–land boundary will attenuate the exchange of air masses and the climate on land will become more continental. However, in coastal regions powerful hurricanes will become more frequent because of the same contrasts.

Yet, there are certain grounds for optimism as well. The abrupt oil crisis of the 1970s, caused by political reasons, prompted the industrialized nations toward the search for alternative sources of energy, especially those not involving combustion of fossil fuel. Despite occasional failures, the technology of nuclear power stations is being improved. The energy output capacity of atomic fission is constantly increasing, and atomic energy may prevail in energy production. A considerable place in energy output should also belong to tidal power stations. Introduction of magnetohydrodynamic generators would be bound to increase the efficiency of thermal and future nuclear power plants. A rhythmic operation of all energy sources could be made possible by powerful energy systems which even today can cover large areas, reducing unproductive energy spending. Energy production thus may become 'cleaner', releasing less CO_2 into the atmosphere and decreasing unnecessary release of heat.

A partial deconcentration of energy sources may occur in the future. The developing nations are gradually upgrading their energy industries. However, the industrialized nations will produce the bulk of the heat in the future, as they do today. It can be expected, therefore, that Europe, North America and Asia will continue to produce most of the artificial heat.

In the more remote future, man will be able to have at his disposal virtually unlimited amounts of energy, which will enhance his capacity for control of the environment. When man succeeds in harnessing the energy of thermonuclear fusion, this will mean a real revolution in energy production. The fuel for thermonuclear fusion – deuterium and lithium – does not have to be mined. The main store of these elements is the world's oceans.

8.4 Currents and transformation on the Earth's climate

The regulating mechanisms of the sea–air system described above will play a major role in the active

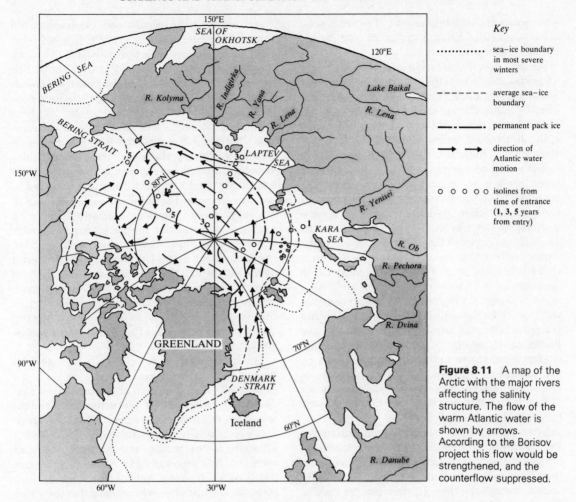

Figure 8.11 A map of the Arctic with the major rivers affecting the salinity structure. The flow of the warm Atlantic water is shown by arrows. According to the Borisov project this flow would be strengthened, and the counterflow suppressed.

influence of man on the Earth's climate. Warming the Arctic has been given the most detailed consideration. We have already pointed out that the sea ice of the Arctic varies in volume and surface area from season to season and from year to year. In Figure 8.11 the boundaries of ice occurrence are shown. Remarkably, these variations correlate closely with the amount of heat received through radiation or heat transport. Budyko (1972) generalized his studies of Arctic glaciation and found that the variation of surface area of sea ice is directly proportional to that of total radiation. A mere 6% increase in the amount of heat received by the Arctic would be sufficient for the complete disappearance of sea ice within 50–70 years. Moreover, this result can be achieved even faster if certain technical measures are taken which have been calculated specifically to increase or maintain the heat flux within the Arctic. In other words, each project for the global regulation of climate is based on an intensification of positive feedback effects such as those that we have considered in studying deglaciation. The engineering solution is aimed at creating a triggering 'heat shock'.

Budyko (1969) and Sellers (1969) have independently shown that a relatively small warming of the Arctic, if well engineered, would lead to a spontaneous subsequent increase of heat with almost no external energy input, once the sea ice has been eliminated in some way. The following considerations are put forward. After the Arctic becomes cleared of ice, its 'dark' water will absorb much more solar radiation than does the present ice. As a result, evaporation will set in, saline convection will begin and the greenhouse effect will be intensified, which will prevent ice formation during winter. Wind activity will cause currents in the ocean, now free of ice, intensifying water exchange with the Atlantic. The north polar ocean, instead of being a source of cold, would become a source of warmth.

We shall now discuss what is understood by the triggering initial impetus with respect to the ice. There are two possible ways of eliminating ice. The first is the action of some substance on the ice. For instance, one could spread dark powder or cultivate algae on the ice surface to increase heat absorption. There are also projects to scatter clouds over the ice or use a monomolecular film to cover ice-free water, or even to increase the amount of heat and moisture in the Arctic by explosion of 'clean' hydrogen bombs or special warming of water by nuclear plants. The second method is to increase the influx of water into the Arctic by deflecting warm currents there. Currents can only be acted upon in some 'bottlenecks'. Such 'bottlenecks' in a figurative sense are straits. Using water exchange through straits, man can compete with nature in determining the regime of a cold water basin. Space would not permit even the listing of all projects of regional and planetary alterations of sea and climatic regimes based on the concept of regulation of water exchange in straits, but among these numerous projects for climatic amelioration, a concept due to P. Borisov (1970) is most widely known.

Since 1958, Borisov has been campaigning for the following scheme for the improvement of climate in the Northern Hemisphere. He suggests the use of powerful pumps to transport Arctic waters through the Bering Strait into the Pacific. The amount of water to be pumped should be initially 145 000 km^3 per year (4–5 times the volume of the annual runoff of all rivers on the Earth) and then raised to 250 000–280 000 km^3 per year. An increased influx of warm Atlantic water is expected to result from this, leading to the formation of a direct stream of warm water through the entire Arctic basin, termed by Borisov the 'Polar Gulf Stream'.

Atlantic water at present takes several years to pass through the Arctic basin counterclockwise (Fig. 8.11). This transport occurs at a depth of 200–700 m, with a water temperature on average of 0.5–0.6°C. Above and below that layer, the water is colder. Sinking to the depths somewhere near the Norwegian Sea, this water does not contribute to warming the Arctic climate. The problem thus is to raise this water to the surface and make it warmer at its very source. These are the objectives of Borisov's project.

According to his calculations, the water at the input of the basin will be at 8.2°C (instead of at 1.9°C as at present). Its low density will not allow the 'Polar Gulf Stream' to sink to the depths. The entire active cross section of the passage between Greenland and Scandinavia will be filled by a flow of Atlantic waters. The countercurrent – the East Greenland Current – will be attenuated or disappear. Borisov has calcualted that the danger of making the Bering Sea or the Pacific Ocean colder is not great since the maximum cooling will not exceed the range of major fluctuations observed in the past 30 years. The author of the project proves the reliability of predicted climatic features by analogy. He compares the paleogeography with the present regime, drawing the conclusion that the Arctic basin has not always been the dead-end of the Atlantic. Before Quaternary time, the two basins had been engaged in active water exchange, which prevented ice formation and had favourable implications for the climate. Thus by creating the direct stream, one could to some extent reconstruct the climate of the past.

The Borisov project is technically feasible with present-day technological skills, but such an endeavor is utterly beyond any economic reason, given the immensity of the water transfer that would be required for the goals of the program. However, socially the Borisov ideas looked very attractive to the public, and his book was well received. The proposal has become famous, and has sparked off several hot debates. A special representative committee was organized in the 1970s to assess the feasibility, effects and economic efficiency of modifying climate by regulating water exchange between the oceans. The conclusions of the committee were negative; moreover, many scientists insisted that the Borisov project, if implemented, would be environmentally damaging. The following simple logical considerations were put forward.

Let us consider, the critics said, what will happen directly after the Arctic basin is cleared of ice and its water temperature rises. The contrasts of temperature will become weaker and the zonal wind activity will attenuate. The subtropical gyre will lose its intensity, and the warm Gulf Stream continuations will not penetrate so far north. The zone of westerlies and northern trade winds will be shifted to the south, and the subtropical gyre will be 'flattened' in the north–south direction. (Recalling Munk's scheme (Fig. 5.6), one can assume that the line $\psi = 0$, which separates the subpolar and subtropical gyres, will be shifted southward.)

Over the surface of the Arctic basin, free of ice, the opponents went on, winds generated by the substantial temperature contrast between the land (Siberia and Greenland) and sea will begin to 'work'. Wind-induced currents will appear that cannot arise now due to the stable ice cover over the larger portion of the water body. The passage between Greenland and Scandinavia is wide enough for a separate circulation to develop there under the effect of non-uniform wind, promoting water exchange in the horizontal plane instead of vertical as it is now. The countercurrent in the form of the East Greenland and Labrador Currents, which according to Borisov should be attenuated, could in fact be greatly increased. It is not to be ruled out that a large portion of the Atlantic waters (which are much colder in fact than Borisov believes) would turn back, never fulfilling its 'heating function'.

Apprehension was also expressed that the artificial flow through the Arctic might eventually shift the thermal Equator. Though there was no paleo-oceanographic evidence of such an effect, it was not ruled out during discussion. Following appreciable weakening of the north–south thermal gradient, the trade wind system would shift southward, and the South Equatorial Current, which now feeds the Gulf Stream system, would begin to do so for the Brazil Current. The latter might become as intensive as the Gulf Stream today.

The pumping of Arctic waters into the Pacific Ocean thus might for a certain period of time improve the heat balance in the Northern Hemisphere. Subsequently, however, one should expect a major outflow of heat from the Northern Atlantic into the Southern Hemisphere. In the Pacific this process would be even faster due to the effects of cold Arctic water. An Arctic isolated from natural heat fluxes would become colder. As a result, negative changes might outweigh the positive ones.

8.5 Regional climate modifications

While projects of climate regulation are fraught with unpredictable effects when applied on a global scale, they can be quite acceptable and economically beneficial on a regional scale. A case in point is the Atlantropa Project described earlier (see Section 2.4). An economic assessment of the benefits of the project (inexpensive, clean power, additional land for development and humidification of Africa) are compared with its negative effects (rebuilding of harbors, replanting of bottom fauna, developing new sea resorts on the Mediterranean, etc.). The effects of the project on other portions of the globe may be insignificant. Even so, one should bear in mind that the Mediterranean raises the salinity of the Atlantic deep water.

Another imaginative project that has won world fame for its author is that of C. L. Riker. His idea is to bring the Gulf Stream and the Labrador Current out of contact by building a 200 mile causeway from Newfoundland along the Grand Banks. In this way, Riker proposed to increase the thermal effect of the Gulf Stream on North America and Europe and at the same time prevent transport of icebergs into the transatlantic shipping lanes. The details of this project – its strengths and weaknesses and the struggle for its approval by the US Congress – are described in a fascinating book by Gaskell (1972). What must be perfectly clear to us here is that a small causeway cannot introduce any fundamental change into the system of currents generated by global processes.

Projects for amelioration of climate of limited regions may prove to be more realistic. One such project, less well known in the West, has been developed by N. M. Budtolaev. It involves modification of the climate of the southern part of the Sea of Okhotsk and the Sea of Japan.

Budtolaev proposes to achieve this modification by altering the water and thermal balances of the Sea of Japan. This is to be done by building dams across the Tsugaru and La Perouse Straits (Fig. 8.12) and reconstructing the Tatarskiy Strait at its narrowest point.

A proposal to build a dam across this part of the strait to link Sakhalin with the mainland – which would have obvious economic benefit – has been made before. In the Tatarskiy Strait, which is approximately 14 km wide, periodic tidal currents are observed transporting waters alternately in northern and southern directions. Budtolaev proposed special openings in the dam with a valve that would only let water through in the northern direction. Building dams in Tsugaru, La Perouse and Tatarskiy Straits is expected to retain in the Sea of Japan the warm waters of a Kuroshio branch flowing around the western coast of southern Sakhalin, move these warm waters into the north-

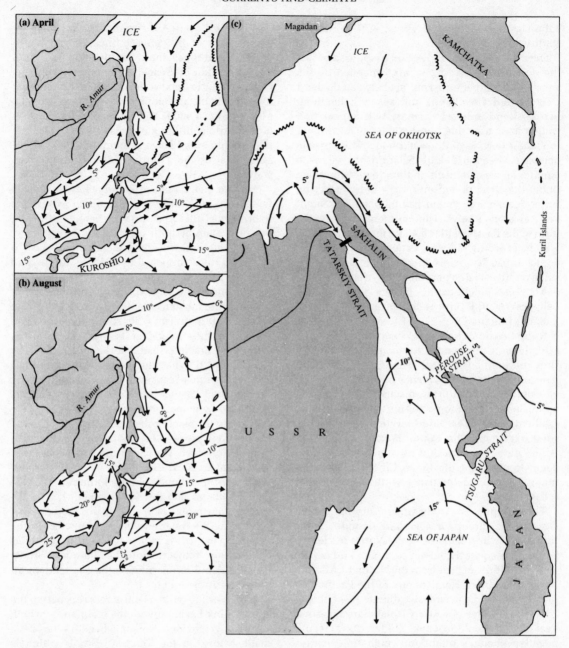

Figure 8.12 The current fields and temperature distributions in the Sea of Okhotsk and the Sea of Japan. (a, b) The present currents and temperatures; (c) after realization of the Budtolaev project.

ern portion of the Tatarskiy Strait and then prevent the influx of cold waters flowing from the north along the Primorskiy coast (Fig. 8.12). This will lead to amelioration of climate in that region and improve navigation in the partially freezing northern Tatarskiy Strait.

The Sea of Okhotsk is marked by an excep-

tionally harsh climate even for these latitudes (50–60°N). The icy breath of the Arctic makes the Sea of Okhotsk clear of ice later than the Bering Sea, which is situated further to the north. The system of currents (Fig. 8.12) is such that the warm waters of the Kuroshio do not penetrate through the Kuril Islands range. The only path for the heat fluxes is through the Tatarskiy Strait. The potential changes of the current system and temperature distribution during the cold season is shown in Figure 8.12.

As with most projects of this kind, the oceanographic and meteorologic aspects have not yet been worked out in detail. Discussions of this project brought up the point that suggested that its implementation is unlikely to result in any noticeable climatic change, because the climate in the region is determined by the general circulation of air in that part of the globe (due to the existence in winter of two powerful centers of atmospheric pressure – the East Siberian high pressure and the Aleutian low pressure – generated by the heat transport of the Kuroshio) rather than by the thermal regime of the Sea of Japan. This system of baric centers gives rise to powerful monsoon-type air fluxes which determine the climate of the region. Naturally, the project could not affect the entire system of circulation, so that changes that could be expected would be moderate and would only affect a small portion of the coast.

This is not true as far as the hydrologic regime of the Tatarskiy Strait is concerned, especially its northern part. A constant flow of warm water through the dam could result in a substantial change of the thermal balance to the positive side, improving navigation conditions during the cold season. A definitive answer to this question can be obtained now by simulating the currents in the strait.

A positive aspect of the project is the possibility of piecemeal implementation involving small specific costs. Indeed, if a dam were built across the strait for transportation, the creation of special openings with valves would involve an outlay imcomparably smaller than the cost of the whole project. This partial implementation will hardly result in major climatic changes, but it will certainly modify the hydrologic regime in the strait.

In general, man's interference with Nature, particularly on a large scale, is dangerous, because the consequences are largely unknown. In 1966 a special report of the US Commission on Weather and Climate Control was issued (*Weather and climate* 1966). The authors of this report, outstanding American specialists on weather and climate, came to the conclusion that with the current level of knowledge of motion in the sea–air system it is impossible to make a reliable prediction of the effects of any 'triggering pulse'. One thing is clear, however: any climatic change, if ever effected, is likely to embrace large territories and, while being favorable in one place, may be deleterious in another.

Soviet geophysicist A. S. Monin was even more definitive. He argued that the atmosphere has a sufficient number of internal mechanisms for resolution of instability. Therefore, the search for 'triggering mechanisms' in the atmosphere and in the ocean is hopeless, for these are not railroads, where by a slight movement of a hand one can switch a heavy freight train into a new direction. Perhaps, these thoughts are a translation into the language of science of the famous words of Goethe:

Nature is always right, always just; mistakes and errors arise from human beings.

References and bibliography

Sources of numeric and graphic information

American practical navigator 1966. Washington, DC: US Government Printing Office.

Bolin, B. 1977. The impact of production and use of energy on the global climate. *Annu. Rev. Energy* **2**, 197–226.

Bogorov, V. G. 1969. *Ocean's life*. Moscow: Znanie (in Russian).

Keeling, C. D. 1980. The oceans and biosphere as future sinks for fossil fuel carbon dioxide. In *Interaction of energy and climate*, W. Bach, J. Pankrath and J. Williams (eds.), 129–47. Dordrecht: Reidel.

Korzun, V. I. (ed.), 1974. *World water balance and water resources of the Earth*. Leningrad: Hydrometeoizdat (in Russian).

Lacombe, M. 1968. *Les energies de la mer*. Paris: Gauthier-Villars (in French).

Lebedev, V. L., T. A. Aizatullin and K. M. Khailov 1974. *Ocean as a dynamical system*. Leningrad: Hydrometeoizdat (in Russian).

Lem, S. 1978. *Solaris*, 2nd edn. New York: Berkeley

Namias, J. 1976. Some statistical and synoptic characteristics associated with El Niño. *J. Phys. Oceanogr.* **6**, 130–8.

Oceanographic atlas of the polar seas, Part II, Arctic, 1959. US Navy Hygrographic Office, Washington, DC, M.O. Publ. No. 703.

Robinson, A. R. 1980. *Dynamics of ocean currents and circulation: Results of POLYMODE and related investigation*. The US POLYMODE Organizing Committee Report, 1–31.

Shamraev, Yu. I. and L. A. Shishkina 1980. *Oceanography*. Leningrad: Hydrometeoizdat (in Russian).

Stepanov, V. N. 1960. General scheme of the surface ocean circulation. In *Physical oceanography*, Vol. 10, No. 1, A. D. Dobrovolsky (ed.), 69–78. Moscow: Izdatelsto AN SSSR (in Russian).

Stepanov, V. N. 1974. *The world ocean: Dynamics and water properties*. Moscow: Znanie (in Russian).

Woodward, W. E. 1972. Instrumentation for measuring ocean currents. *ISA Trans* **11**, 327–30.

Wooster, W. S. (ed.) 1970. *Scientific exploration of the South Pacific*. 45–6. Washington, DC: National Academy of Sciences.

Zubov, N. N. 1947. *Dynamical oceanography*. Leningrad: Hydrometeoizdat (in Russian).

Popular science readings

Burt, W. V. and S. A. Kulm 1966. Oceanography. In *The encyclopedia of oceanography*, R. W. Fairbridge (ed.), 610–13. Stroudsburg, PA: Dowden, Hutchison and Ross.

Charlier, R. H. and B. L. Gordon 1978. *Ocean resources: An introduction to economic oceanography*. Washington, DC: University Press of America.

Charlier, R. H., B. L. Gordon and J. Gordon 1980. *Marine science and technology. An introduction to oceanography*. Washington, DC: University Press of America.

Cromie, W. J. 1962. *Exploring the secrets of the sea*. Englewood Cliffs, NJ: Prentice-Hall.

Deacon, G. E. 1962. *Seas, maps, and men*. Garden City, NY: Doubleday.

Deacon, M. 1971. *Scientists and the sea, 1650–1900: A study of marine science*. New York: Academic Press.

Fedorov, K. N. 1977. Hidden dynamics of Neptune's powerhouse. *Unesco Courier* **30**, 24–7.

Flohn, H. 1966. Energy budget of the Earth's surface. In *The encyclopedia of oceanography*, R. W. Fairbridge (ed.), 250–6. Stroudsburg, PA: Dowden, Hutchison and Ross.

Gaskell, T. F. 1972. *The Gulf Stream*. New York: John Day.

Gribbin, J. 1979. Climatic impact of Soviet river diversions. *New Scientist* **84**, 762–5.

Gross, M. G. 1972. *Oceanography: A view of the Earth*. Englewood Cliffs, NJ: Prentice-Hall.

Hammond, A. L. 1974. Undersea storms: experiment in the Atlantic. *Science* **185**, 244–7.

Hartline, B. K. 1979. POLYMODE: exploring the undersea weather. *Science* **205**, 571–3.

Johnson, J. W. 1966. Ocean currents, Introduction. In *The encyclopedia of oceanography*, R. W. Fairbridge (ed.), 587–90. Stroudsburg, PA: Dowden, Hutchison and Ross.

Oceanus, 1976. Special issue on 'Ocean eddies'. *Oceanus* **19** (3).

Oceanus, 1981. Special issue on 'Oceanography from space'. *Oceanus* **23** (3).

Romanovsky, V., C. Francis-Boeuf and J. Bourcart 1953. *La mer*. Paris: Librairie Larousse (in French).

Ross, D. A. 1980. *Opportunities and uses of the ocean*. New York: Springer-Verlag.

Stewart, R. W. 1967. The atmosphere and the ocean. In *Oceanography readings from Scientific American*, J. R. Moor (ed.), 25–44. San Francisco: W. H. Freeman.

Stommel, M. 1967. The circulation of the abyss. In *Oceanography readings from Scientific American*, J. R. Moor (ed.), 77–81. San Francisco: W. H. Freeman.

Tolmazin, D. 1976. *Ocean in movement*. Leningrad: Hydrometeoizdat (in Russian).

Understanding climatic changes, A program for action, 1975. Washington, DC: Committee for the Global Atmospheric Research Program, National Academy of Sciences.

Vetter, C. (ed.) 1973. *Oceanography, The last frontier*. New York: Basic Books.

Weather and climate: Modification problems and prospects, 1966. NAS/NRC Publ. No. 1350, Washington DC.

Wilmot, P. D. and A. Slingerland 1977. Technology assessment and the oceans. *Proc. Int. Conf. on Technology Assessment*, Monaco, Oct. 23–30, 1975. Surrey, England: IPC Science and Technology Press.

Wilson, K. L. (ed.) 1971. *Inadvertent climate modification*. Cambridge, MA: MIT Press.

Textbooks

Introductory and intermediate

Anikouchine, W. A. and R. W. Sternberg 1981. *The world ocean. An introduction to oceanography*, 2nd edn. Englewood Cliffs, NJ: Prentice-Hall.

Drake, C., J. Imbrie, J. A. Knauss and K. K. Turekian 1978. *Oceanography*. New York: Holt, Rinehart and Winston.

Knauss, J. A. 1978. *Physical oceanography*. Englewood Cliffs, NJ: Prentice-Hall.

McCormick, J. M. and J. V. Thiruvathukal 1981. *Elements of oceanography*, 2nd edn. Lavallette, NJ: Saunders.

Perry, A. H. and J. M. Walker 1977. *The ocean–atmosphere system*. London: Longman.

Pickard, G. L. 1975. *Descriptive physical oceanography*, 2nd edn. Oxford: Pergamon Press.

Pipkin, B. W., D. S. Gorsline, R. E. Casey and D. E. Hammond 1977. *Laboratory exercises in oceanography*. San Francisco: W. H. Freeman.

Ross, D. A. 1970. *Introduction to oceanography*. New York: Appleton-Century-Crofts.

Stowe, K. S. 1979. *Ocean science*. New York: Wiley.

Strahler, A. N. 1971. *The earth science*. New York: Harper and Row.

Weyl, P. K. 1970. Oceanography. An introduction to the marine environment. New York: Wiley.

Advanced

Dietrich, G. and K. Kalle 1963. *General oceanography – An introduction*. New York: Wiley.

Ivanoff, A. 1972. *Introduction à l'oceanographie*, Vol 1. Paris: Librairie Vuibert (in French).

Krummel, O. 1907, 1911. *Handbuch der' ozeanographie*, Vol. I (1907), Vol. II (1911). Stuttgart: Engelhorn (in German).

Lacombe, M. 1965. *Cours d'oceanographie physique*. Paris: Gauthier-Villars (in French).

Neumann, G. 1968. *Ocean currents*. Amsterdam: Elsevier.

Neumann, G. and W. J. Pierson, Jr 1966. *Principles of physical oceanography*. Englewood Cliffs, NJ: Prentice-Hall.

Pond, S., and G. L. Pickard 1978. *Introductory physical oceanography*. Oxford: Pergamon Press.

Sverdrup, M. U., M. W. Johnson and R. M. Fleming 1942. *The oceans, their physics, chemistry and general biology*. Englewood Cliffs, NJ: Prentice-Hall.

Von Arx, W. S. 1962. *An introduction to physical oceanography*. New York: Addison-Wesley.

Von Schwind, J. J. 1980. *Geophysical fluid dynamics for oceanographers*. Englewood Cliffs, NJ: Prentice-Hall.

Monographs

Bach, W., J. Pankrath and J. Williams 1980 (eds.). *Interaction of energy and climate*. Dordrecht: D. Reidel.

Belyaev, V. I. 1975. *Control of natural environment*. Kiev: Naukova Dumka.

Borisov, P. M. 1970. *Can man change the climate?* Moskow: Nauka (in Russian).

Budyko, M. I. 1972. *Man's impact on climate*. Leningrad: Hydrometeoizdat (in Russian).

Budyko, M. I. 1974. *Climate and life*. New York: Academic Press.

Chorley, R. J. and B. A. Kennedy 1971. *Physical geography: A system approach*. Englewood Cliffs, NJ: Prentice-Hall.

Defant, A. 1961. *Physical oceanography*, Vol. 1. Oxford: Pergamon Press.

Felsenbaum, A. I. 1960. *Fundamentals and methods of calculations of steady sea currents*. Moscow: USSR Academy Publ. House (in Russian).

Humboldt, A. von 1814. *Voyage aux regions equinoxiales de nouveau continent, fait en 1799–1804*, 3 vols. Paris.

Kamenkovich, V. M. 1977. *Fundamentals of ocean dynamics*. Amsterdam: Elsevier.

Leonov, A. K. 1966. *Regional oceanography*. Leningrad: Hydrometeoizdat (in Russian).

Mamaev, O. I. 1962. *No-motion reference surface in The world ocean*. Moscow: The Moscow Univ. Press (in Russian).

Mamaev, O. I. 1975. *Temperature–salinity analysis of world ocean waters*. Amsterdam: Elsevier.

Monin, A. C. 1972. *Weather forecasting as a problem in physics*. Boston: MIT Press.

Monin, A. S., V. M. Kamenkovich and V. G. Kort 1977. *Variability of the oceans*. New York: Wiley Interscience.

Monin, A. S. and A. M. Yaglom 1971. *Statistical fluid mechanics*, Vol. 1. Cambridge, MA: MIT Press.

Shuleikin, V. V. 1953. *Sea physics*. Moscow: USSR Acad. of Sci. Press (in Russian).

Starr, V. P. 1968. *Physics of negative viscosity phenomena*. New York: McGraw-Hill.

Stommel, M. 1958. *The Gulf Stream. A physical and dynamical description*. Boston, MA: Cambridge University Press.

Tennekes, H. and J. L. Lumley 1972. *A first course in turbulence*. Cambridge, MA: MIT Press.

Articles

Anouchi, A., D. Schiff and S. Nayes 1974. New approaches to ocean current measurements. *ISA Trans* **13**, 303–10.

Apel, J. R. 1976. Ocean science from space. *EOS, Trans Am. Geophys. Union* **57**, 612–24.

Baker, D. J. 1981. Ocean instruments and experiment design. In *Evolution of physical oceanography*, B. A. Warren and C. Wunch (eds.), 396–433. Cambridge, MA: MIT Press.

Bjerknes, J. 1966. Survey of El Niño 1957–58 in its relation to tropical Pacific meteorology. *Bull. Int. Am. Trop. Tuna Comm.* **12**, 25–86.

Bryan, K. and M. D. Cox 1972. The circulation of the world ocean: A numerical study. *J. Phys. Oceanogr.* **2**, 326–36.

Budyko, M. I. 1969. The effect of solar radiation variations on the climate of the Earth. *Tellus* **21**, 611–19.

Bulatov, R. P., M. S. Barash, V. N. Ivanenkov and Yu. Yu. Marti 1977. *The Atlantic Ocean*. Moscow: Mysl (in Russian).

Burkov, V. A. 1980. *General circulation of the world ocean*. Leningrad: Hydrometeoizdat.

Carrier, G. F. and A. R. Robinson 1962. On the theory of the wind-driven ocean circulation. *J. Fluid Mech.* **12** (1), 49–80.

Charney, J. G. 1960. Nonlinear theory of a wind-driven homogeneous layer near the equator. *Deep-sea Res.* **6** (4), 303–10.

CLIMAP 1976. The surface of the ice-age Earth. *Science* **191**, 1131–7.

Cox, M. D. 1975. A baroclinic numerical model of the world ocean: preliminary results. In *Numerical models of ocean circulation*, 107–20. Washington, DC: National Academy of Sciences.

Deep-Sea Research 1960 **6** (4).

Defant, A. 1936. Die Troposphare, Deutsche Atl. Exped. 'Meteor', 1925–1927. *Wiss. Erg.* **6** (1), 289–411.

Dobrovolsky, A. D., B. S. Zalogin and A. N. Kosarev 1974. Technology of oceanographic research. In *Oceanology*, Vol. 1, A. P. Kapitsa (ed.), 96–127. Boston: G. K. Hall.

Felsenbaum, A. I. 1970. Dynamics of ocean currents. In *Hydrodynamics*, L. I. Sedov (ed.), 97–338. Moscow: Viniti (in Russian).

Fofonoff, N. P. 1954. Steady flow in a frictionless homogeneous ocean. *J. Marine Res.* **13** (3), 254–62.

Foster, T. D. and E. C. Carmack 1976. Temperature and salinity structure in the Weddell Sea. *J. Phys. Oceanogr.* **6**, 36–44.

Fuglister, F. C. 1960. *Atlantic Ocean atlas of temperature and salinity profiles and data from the International Geophysical Year of 1957–1958.* Woods Hole oceanographic Institution Atlas Series No. 1.

Fuglister, F. C. 1972. Cyclonic rings formed by the Gulf Stream, 1965–66. In *Studies in physical oceanography*, A. Gordon (ed.), 137–66. New York: Gordon and Breach.

Gates, W. L. 1976. Modeling the ice-age climate. *Science* **191**, 1138–44.

Hall, A. D. and R. E. Fagen 1968. Definition of system. In *Modern systems research for the behavioral scientist*, W. Buckley (ed.). Chicago: Aldine.

Helland-Hansen, B. 1976. Nogen hydrografishe metoder. In *Forhand lingerved de 16 Skandinaviske naturforskerermote*, 357–9.

Huang, N. E. 1979. New development in satellite oceanography and current measurements. *Papers in Oceanogr.* **17** (7), 1558–68.

Ilyin, A. M. and V. M. Kamenkovich 1964. On the structure of the boundary layer in the two-dimensional current theory. *Oceanology* **4** (5), 756–69 (Russian, English translation).

Imbrie, J. and J. Z. Imbrie 1980. Modeling the climatic response to orbital variations, *Science* **207**, 943–53.

Isaacs, J. D. and W. R. Schmitt 1969. Simulation of marine productivity with waste heat and mechanical power. *Couseil* **33**, 71–87.

Iselin, C. O. 1936. A study of the circulation of the western North Atlantic. *Pap. Phys. Oceanogr. Meteor.* **4** (4).

Kamenkovich, V. M. and A. S. Monin (eds.) 1978. *Oceanic physics*, Vol. 1: *Oceanic hydrophysics*, Vol. 2: *Oceanic hydrodynamics*. Moscow: Nauka (in Russian).

Kolmogorov, A. 1941. The local structure of turbulence in incompressible viscous fluid for very large Reynolds numbers. *Dokl. Akad., Nauk SSSR* **30**, 301.

Kort, V. G. 1969. The main scientific results of the exped-

ition on the RV Akademic Kurchatov (fifth voyage). *Oceanology* **9**, 744–51.

Kuftarkov, Yu. M., B. A. Nelepo and A. D. Fedorovsky 1978. On cold temperature skin-layer of the ocean. *Izv. USSR Acad. Sci., Atmos. Ocean. Phys.* **14**, 88–93 (English translation).

Lamb, H. M. 1969. Climate fluctuation. In *World survey of climatology*, M. Flohn (ed.), 2.173–249. Amsterdam: Elsevier.

Land, T. 1974. Exchanging ocean data like weather reports. *Sea frontiers* **20**, 13–15.

Latun, V. S. 1971. Upwelling. *The earth and the universe* **9**, 37–41 (in Russian).

Levy, E. M., M. Ehrhardt, D. Kohuke, E. Sobtchenko, T. Suzuoki and A. Tokuhiro 1981. *Global oil pollution.* Paris: IOC Unesco.

Lineikin, P. S. 1955. On the determination of depth of a baroclinic layer in the sea. *(Doklady) C. R. Acad. Sci, USSR* **101** (3), 461–3.

Milankovitch, M. M. 1941. Kanon der Erdbestrahlung und seine Anwendung auf das Eiszeitenproblem. *Acad. Royale, Ser. B Edition, Spes, 133*, Sect. Sci. Mathem. Natur. 44. (Translated by the Israel Program for Scientific Translations, Jerusalem, 1969.)

Montgomery, R. V. and E. Palmén 1940. Contribution to the question of the equatorial counter current. *J. Marine Res.* **3**, 112–33.

Moore, D. W. 1963. Rossby waves in ocean circulation. *Deep-Sea Res.* **10**, 735–47.

Munk, W. H. 1950. On wind-driven circulation. *J. Meteorol.* **7** (2), 79–93.

Munk, W. H. and C. Wunch 1979. Ocean acoustic tomography: a scheme for large scale monitoring. *Deep-Sea Res.* **26**, 123–61.

Neumann, G. 1956. Notes on the horizontal circulation of ocean currents. *Bull. Am. Meteorol. Soc.* **37** (3), 96–100.

Obukhov, A. M. 1941. On energy distribution in a spectrum of turbulent flow. *Izv. Geograph. Geophys. Ser.* 5, No. 4–5, 453–66 (in Russian).

Okubo, A. and R. V. Ozmidov 1970. Empirical dependence of the coefficient of horizontal turbulent diffusion in the ocean on the scale of the phenomena in question. *Izv. Atm. Ocean. Phys.* **5**, 308–9.

Ozmidov, P. V. 1965a. Energy distribution between oceanic motions of different scales. *Atm. Ocean. Phys.* **1**, 257–61.

Ozmidov, P. V. 1965b. On the turbulent exchange in a stably stratified ocean. *Atm. Ocean. Phys.* **1**, 493–7 (English translation).

Parsons, A. T. 1969. A two layer model of Gulf Stream separation. *J. Fluid Mech.* **27** (2), 179–90.

Reid, J. L., W. D. Nowlin and W. C. Patzert 1977. On the characteristics and circulation of the Southwestern Atlantic Ocean. *J. Phys. Oceanogr.* **7**, 62–91.

Reid, R. O. 1948. A model of the vertical structure of mass in equatorial wind-driven currents of a baroclinic ocean. *J. Mar. Res.* **7**, 304–12.

Rhines, P. 1976. The dynamics of unsteady currents. In *The sea*, Vol. 6: *Marine Modeling*. E. D. Goldberg, I. N. McCave, J. J. O'Brien and J. M. Steele (eds.). New York: Wiley.

Richardson, L. F. 1926. Atmospheric diffusion shown on a distance–neighbour graph. *Proc. R. Soc.* **4** (110), 709–37.

Richardson, P. L. 1980. Gulf Stream ring trajectories. *J. Phys. Oceanogr.* **10**, 90–104.

Ring Group 1981. Gulf Stream rings, their physics, chemistry and biology. *Science* **212**, 1091–100.

Robinson, A. R. 1980. *Dynamics of ocean currents and circulation: results of POLYMODE and related investigations.* Boston: US POLYMODE Organizing Committee, MIT.

Robinson, A. R. 1982. Eddies in marine science. In *Woods Hole Oceanographic Institution 50 year Anniversary Volume.* Cambridge, MA: Woods Hole Oceanographic Institution.

Robinson, A. R., D. E. Harrison and D. B. Maidvogel 1979. Mesoscale eddies and general ocean circulation models. *Dyn. Atmos. Oceans* **3**, 143–80.

Rossby, T. 1981. Eddies and the general circulation. In *Woods Hole Oceanographic Institution 50 year Anniversary Volume.* Cambridge, MA: Woods Hole Oceanographic Institution.

Rossby, H. T., A. Voorhis and D. Webb 1975. A quasi-Lagrangian study of mid-ocean variability using long range SOFAR floats. *J. Mar. Res.* **33**, 355–82.

Ruddiman W. F. and A. McIntyre 1979. Warmth of the subpolar North Atlantic Ocean during Northern Hemisphere ice-sheet growth. *Science* **204**, 173–5.

Ruddiman, W. F. and A. McIntyre 1981. Oceanic mechanisms of amplification of the 23 000 year ice-volume cycle. *Science* **213**, 617–27.

Sellers, W. D. 1969. A global climatic model based on the energy balance of the Earth–atmosphere system. *J. Appl. Meteor.* **8**, 392–411.

Shtockman, V. B. 1946a. Equations for a field of total flow induced by the wind in a non-homogeneous sea. (*Doklady*) *C.R. Acad. Sci., USSR* **54** (5), 407–10.

Shtockman, V. B. 1946b. A theoretical explanation of certain peculiarities of the meridional profile of the surface of the Pacific Ocean. (*Doklady*) *C.R. Acad. Sci., USSR* **53**, 323–4.

Shtockman, V. B. 1948a. Influence of bottom relief on the direction of mean transport induced by wind and mass distribution in non-homogeneous ocean. (*Doklady*) *C.R. Acad. Sci., USSR* **59** (5), 889–92.

Shtockman, V. B. 1984b. *Equatorial countercurrents in the oceans.* Leningrad: Hydrometeoizdat (in Russian).

Shtockman, V. B. 1952. On steady-state currents and density distribution in a central cross-section of a closed sea of elongated form. *Izv. Geophys. Ser.* **6**, 57–72 (in Russian).

Simonov, A. I. 1979. Monitoring of chemical pollution at sea. In *Problems of research and exploration of the World Ocean*, A. I. Yoznesensky (ed.), 93–108. Leningrad: Sudostroenie.

Stommel, H. 1948. The westward intensification of wind-driven ocean currents. *Trans. Am. Geophys. Union* **29** (2), 202–6.

Stommel, H. 1957. A survey of ocean currents theory. *Deep-Sea Res.* **4**, 149–84.

Stommel, H. 1960. Wind-drift near equator. *Deep-Sea Res.* **6** (4), 292–302.

Stommel, H. and A. B. Arons 1960. On the abyssal circulation of the World Ocean – I. Stationary planetary flow patterns on a sphere. *Deep-Sea Res.* **6**, 140–54; II. An idealized model of the circulation pattern and amplitude in oceanic basins. *Deep-Sea Res.* **6**, 217–33.

Sverdrup, H. 1940. Lateral mixing in the deep water of the South Atlantic Ocean. *J. Mar. Res.* **2** (3).

Sverdrup, H. V. 1947. Wind-driven currents in a baroclinic ocean, with application to the equatorial currents of eastern Pacific Ocean. *Proc. Nat. Acad. Sci.* **33**, 318–26.

Swallow, J. C. and L. V. Worthington 1961. An observation of a deep countercurrent in the Western North Atlantic. *Deep-Sea Res.* **8** (1), 1–19.

Szekielda, K. 1976. Spacecraft oceanography. *Oceanogr. Mar. Biol. Annu. Rev.*, **14**, 99–166.

Tolmazin, D. 1962. On the problem of positions of zero reference surface in the ocean. *Oceanology* **2** (5), 815–23 (Russian, English abstract).

Tolmazin, D. 1972. Features of horizontal turbulence in the littoral zone of the ocean. *Izv. Atmos. Ocean. Phys.* **8** (3), 194–6.

Tolmazin, D. M. 1974. Sea straits. In *Oceanology*, A. P. Kapitsa (ed.). Boston, MA: Hall.

Tolmazin, D. and V. Shneidman 1968. Calculations of integral circulation and currents in the northwestern part of the Black Sea. *Izv. Atmos. Ocean. Phys.* **4**, 361–6.

Veronis, G. 1966. Wind-driven ocean circulation – Part 2. Numerical solutions of the non-linear problem. *Deep-Sea Res.* **13**, 30–5.

Veronis G. 1973. Model of world ocean circulation: I. wind-driven, two layer. *J. Mar. Res.* **31**, 228–88.

Veronis, G. 1981. Dynamics of large-scale ocean circulation. In *Evolution of physical oceanography*, B. A. Warren and C. Wunch (eds.), 140–83. Cambridge, MA: MIT Press.

Vulfson, V. I. 1970. The problem of free oxygen resources. *Oceanology* **10** (3), 295–301.

Warren, B. 1966. Oceanic circulation. In *The encyclopedia of oceanography*, R. W. Fairbridge (ed.), 590–7. Stroudsburg, PA: Dowden, Hutchison and Ross.

Warren, B. A. 1974. Deep flow in the Madagascar and Mascarene Basins. *Deep-Sea Res.* **21**, 1–21.

Warren, B. A. 1981. Deep circulation of the World Ocean. In *Evolution of physical oceanography*, B. A. Warren and C. Wunch (eds.), 6–41. Cambridge, MA: MIT Press.

Weyl, P. K. 1968. The role of the oceans in climatic change: A theory of the ice ages. *Meteor. Monographs* **8** (30) 37–62.

Welander, P. 1959. Sea–air coupling. In *Proceedings of the first Oceanographic congress*, M. Sears (ed.), 35–41. New York: Wiley.

Worthington, L. V. 1954. Three detailed cross-sections of the Gulf Stream. *Tellus* **6**, 116–28.

Worthington, L. V. 1970. The Norwegian Sea as a mediterranean basin. *Deep-Sea Res.* **17**, 77–84.

Worthington, L. V. 1981. The water masses of the World Ocean: Some results of a fine-scale census. In *Evolution of physical oceanography*, B. A. Warren and C. Wunch (eds.). Cambridge, MA: MIT Press.

Wust, G. 1936. Kuroshio und Golfstrom. *Veroff. d. Inst. an d. Univ. Berlin*, N. F. Reihe A: Geogr-naturwiss., Heft 29.

Wust, G. 1964. The major deep-sea expeditions and research vessels, 1873–1960. In *Progress in oceanography*, Vol. 2, 1–52. New York: Pergamon Press.

Zenkovich, L. A. and Yu. Yu. Marti 1970. On aquaculture farms. *Priroda (Nature)* **4**, 52–8.

Index

References to text sections are given in **bold** type, to text figures in *italics*, and to tables as, e.g. 'Table 7.2.'